Life-Cycle Greenhouse-Gas Emissions Inventory
For Fischer-Tropsch Fuels

Prepared for

U.S. Department of Energy
National Energy Technology Laboratory

Prepared by

Energy and Environmental Solutions, LLC

John J. Marano
Jared P. Ciferno

June 2001

ACKNOWLEDGEMENTS

The authors would like to express their appreciation to all individuals who contributed to the successful completion of this project and the preparation of this report. This includes Dr. Shelby Rogers, Dr. Rodney Geisbrecht, and Dr. Michael Nowak of the U.S. DOE for their insightful comments and support through this endeavor. It also includes Ms. Lynn Manfredo and Dr. Victor Gorokhov of SAIC, both of whom provided data that was included in our analysis. We would also like to thank all those who agreed to provide peer review for this report, including Dr. Gerald Choi and Dr. Paul Worhach, Nexant; Dr. David Gray, Mitretek; Dr. Howard McIlvried, SAIC; Dr. John Shen; DOE; and Ms. Pamela Spath, NREL. Peer review is a time consuming process, which is rarely tangibly rewarded, but for which we are sincerely grateful. Finally, thanks are given to Ms. Colleen Hitchings for her tireless assistance in preparing this document.

DISCLAIMER

EXECUTIVE SUMMARY

This report discusses the development of greenhouse gas (GHG) emissions estimates for the production of Fischer-Tropsch (FT) derived fuels (in particular, FT diesel), makes comparisons of these estimates to reported literature values for petroleum-derived diesel, and outlines strategies for substantially reducing these emissions. This report is the product of the first phase of a comprehensive assessment being conducted by Energy and Environmental Solutions (E[2]S), LLC, for the National Energy Technology Center (NETL) to characterize the impact, both short and long term, of FT fuel production on the environment and on human health and well-being.

This study involved the development of GHG inventories for a number of conceptual FT process designs. It also included the development of preliminary estimates for criteria pollutant emissions. The next phase of this assessment will address life-cycle improvements for FT fuels by targeting specific process changes aimed at reducing GHG emissions. Preliminary results have identified promising reduction strategies and these estimates have been included in this document. Future research will be focused on expanding the current emissions inventory to include a broader range of multimedia emissions of interest to NETL programs, and on performing economic analyses corresponding to the new low-emission FT process designs developed.

Baseline GHG Inventory

The objective of this study was to conduct a *full* life-cycle inventory (LCI) of greenhouse gas emissions for synthetic fuels produced using the FT process. As shown below, the LCI is based on a *"cradle-to-grave"* approach and includes data identification, collection and estimation of GHG emissions from *upstream extraction/production, conversion/refining, transportation/distribution, and end-use combustion of FT fuels* derived from three types of feedstocks: coal, biomass and natural gas.

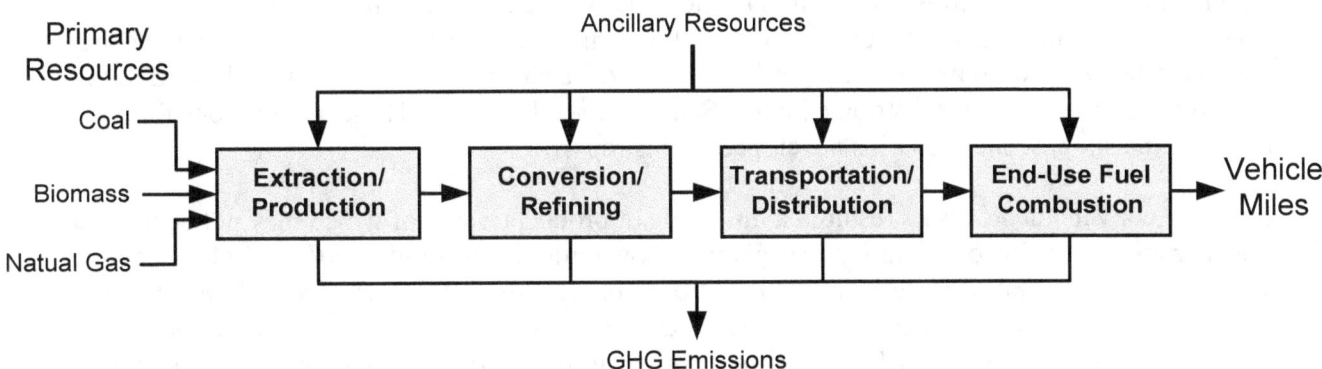

The material and energy balances used for this analysis are from conceptual process designs developed for DOE in the 1990s for coal liquefaction and gas-to-liquid (GTL) plants[1].

---

1. Bechtel, Inc. *Baseline Design /Economics for Advanced Fischer-Tropsch Technology* (various reports), DOE Contract No. DE-AC22-91PC90027 (1993-1998).

Background: The analysis presented in this report is limited to a LCI of airborne emissions produced along the FT fuel product life cycle. It is not a *complete* inventory of all emissions, though it could be used as a starting point for one, since it lays out *a formal methodology for conducting an analysis for FT derived fuels*. The impact of various greenhouse gases has been considered in relative terms by converting all GHG emissions to a CO_2 equivalency basis. The LCI is based on earlier FT plant designs, and no effort has been made to improve on these conceptual designs.

The greenhouse gases considered are CO_2 (carbon dioxide) from syngas production, FT synthesis, fossil-fuel combustion along the life-cycle, and venting from natural gas production; CH_4 (methane) from fugitive plant and pipeline emissions, incomplete combustion or incineration (gas flaring), and coalbed methane releases; and N_2O (nitrous oxide) from fuel combustion and the cultivation of biomass. The weighting factors for CH_4 and N_2O used in the CO_2 equivalency calculations are 21 and 310, respectively. Data were also compiled, where possible, for emissions of criteria pollutants (CP): CO (carbon monoxide), NOx (nitrogen oxides), SOx (sulfur oxides), VOC (Volatile Organic Compounds), and PM (Particulate Matter). Normally, these emissions are not included in CO_2 equivalency calculations, because the mechanism of their participation in global warming is not fully understood. For the FT conversion process, a checklist of air toxics sources has also been prepared.

Assumptions relative to the geography of the product supply chain (*fuel chain*) are critical when comparing life-cycle emissions estimates. The U.S. Midwest (southern Illinois) has been chosen as a reasonable location for the future siting of coal liquefaction plants, as well as biomass conversion plants. A Wyoming location was also chosen for a second coal scenario based on the conversion of subbituminous coal. For these scenarios, it was assumed that the FT diesel fuel is supplied to an area in the vicinity of Chicago, IL by pipeline and tank truck. Three locations were considered for siting a GTL plant: southern Illinois, Venezuela, and Alaska. The southern Illinois location has been included to allow direct comparison between coal, biomass and natural gas scenarios. For Venezuela, it is assumed that FT syncrude is transported to the U.S. Gulf Coast by tanker and pipelined to the U.S. Midwest, where it is refined and blended into transportation diesel fuel near Chicago. It is assumed that GTL deployment on the North Slope of Alaska results in a syncrude that is transported via the Trans-Alaska pipeline to Valdez, transferred to a tanker, and shipped to the U.S. West Coast, where it is distributed in the San Francisco Bay area. These assumptions form the basis for the six baseline scenarios developed in this report.

Since FT conversion processes result in a multitude of products, some of which may not be used in transportation, careful consideration was given to how emissions should be *allocated* between the various products. *For this study, emissions from conversion/refining, and all other upstream operations have been allocated between LPG, gasoline and distillate fuel products based on the ratio of their energy content (LHV-basis) to the energy content of all products.* It is unlikely that more complicated procedures would result in substantially different results, since the energy densities of these liquid fuels are similar. However, this procedure was not considered appropriate when electric power was produced as a major by-product of FT production. *Emissions are allocated to power based on the energy content of the fuel used in the electrical conversion device* (gas or steam turbine); that is, the energy content of the electrical power is divided by turbine efficiency when determining the share of emissions to be allocated to this power. This is similar to the procedure used when calculating the *thermal efficiency* of co-generation (power and steam) processes. *The allocation procedure used for fuels and power co-production has a significant effect on the reported emissions.* Further work is needed to validate any benefits of co-production.

The basis for the full FT fuel chain GHG emissions estimates reported here is vehicle-miles driven. This is the appropriate unit of measure for most, but not all, comparisons. Fuel economies in miles-per-gallon (mpg) are from a recent analysis conducted by Argonne National Laboratory (ANL)[2]. This analysis considered a wide range of conventional, advanced, and electric hybrid gasoline and diesel powered vehicles. Since the emissions estimates will change based on the fuel economy used for the comparison, the calculations have been incorporated into a spreadsheet to facilitate analysis of various alternatives with different mpg ratings. The values presented here are for sport utility vehicle (SUV) conversion from conventional gasoline engines to conventional and advanced diesel engines. The average fuel economy for gasoline-powered SUVs is 20 mpg, and for light-duty diesel-powered vehicles it is about 39 mpg. In similar applications, diesel engines are 33% more efficient than gasoline engines. Therefore, converting all SUVs powered by gasoline to diesel would result in a fuel economy increase to 26.6 mpg. Fuel composition also plays a critical role in determining fuel economy. Substituting FT diesel for petroleum diesel in SUVs would result in a decrease in fuel economy from 26.6 to about 24.4 mpg, an 8% decrease. This is a result of the inherent lower energy density per gallon of FT diesel relative to conventional petroleum diesel.

---

2. "Well-to-Wheel Efficiency Analysis Sees Direct-Hydrogen Fuel Cells, Advanced Diesel Hybrids Comparable," *Hart's Gas-to-Liquids News*, April 1999.

Results: As part of this analysis, a large number of FT fuel-chain options were considered, including primary feedstock, production/extraction location and method, FT catalyst and upgrading, FT product slate, co-production of power, transportation method and distances, and end-use location.

FT Fuel-Chain Options

Feedstocks	Production/ Extraction	Conversion/ Refining	Transportation/ Distribution
Coals: • Illinois #6 – bituminous • Powder River Basin – subbituminous	Underground Mining: • S. Illinois Surface Mining: • S. Illinois • Wyoming	FT Conversion: • Iron Catalyst FT Upgrading: • Max Distillate • Max Naphtha • Chemicals	Mine-Mouth FT Plant: • S.IL to Chicago – Pipeline & Tank Truck • Wyo. to Chicago – Pipeline & Tank Truck
Biomass: • Maplewood	Plantation Crop: • S. Illinois	FT Conversion: • Iron Catalyst FT Upgrading: • Fuels & Power	FT Plant near Plantation: S.IL to Chicago – Pipeline & Tank Truck
Natural Gas: • Pipeline Gas • Associated Gas	Pipeline Gas: • S.Illinois Associated Gas: • Venezuela • Alaska North Slope	FT Conversion: • Cobalt Catalyst FT Upgrading: • Max Distillate • Min Upgrading • Fuels & Power	S.IL & Wellhead FT Plant: • S.IL to Chicago – Pipeline & Tank Truck • Venezuela to Chicago – Tanker, Pipeline & Tank Truck • Alaska to Chicago – Pipeline, Tanker & Tank Truck

The only end-use option considered here was diesel-powered SUVs, though cases can be quickly compiled for other applications using the information presented in this report.

A summary of selected results from the GHG emissions inventory developed for FT diesel is given below. Also included are literature estimates for petroleum-derived diesels from imported Arab Light crude oil and a partially upgraded Venezuelan syncrude[3]. Literature data was also used to estimate emissions for Alaska North Slope (ANS) and Wyoming crude oils of direct interest to this study.

Full Life-Cycle GHG Emissions for FT & Petroleum Diesel Scenarios
(g CO_2-eq/mile in SUV)

Feedstock	Extraction/ Production	Conversion/ Refining	Transport./ Distribution	End Use Combustion	Total Fuel Chain
IL #6 Coal baseline	26	543	1	368	939
- in advanced diesel*	23	472	1	320	816
Wyoming Coal	7	585	2	368	962
Plantation Biomass	-969	703	1	368	104
Pipeline Natural Gas	71	121	1	368	562
Venezuelan Assoc. Gas	51	212	12	368	643
- with flaring credit*	-527	212	12	368	65
ANS Associated Gas	51	212	21	368	652
Wyoming Sweet Crude Oil	23	74	8	363	468
Arab Light Crude Oil	35	81	26	367	509
ANS Crude Oil	28	101	14	378	522
Venezuelan Syncrude	32	143	10	390	574

*selected cases from sensitivity analysis.

The figure given on the following page compares graphically the GHG emissions for those baseline scenarios listed above, which produce diesel fuel for the Chicago market.

The results in this table and figure illustrate a number of interesting points. Emissions from transportation (1 to 26 g CO_2-eq/mile) correlate with the distance the fuel or feedstock is moved to market. Thus, in a carbon-constrained world it may not make environmental sense to move oil (or any other commodity) halfway around the world. Transportation emissions are low for domestic coal and biomass-based FT conversion due to the close vicinity of the coal field or plantation and the FT plant to the fuel market (Chicago). The end-use combustion emissions for FT diesel have been assumed constant (368 g/mile in conventional diesel and 320 g/mile in advanced diesel), since the different feedstocks are being refined to produce similar quality products. Emissions for petroleum-derived diesel vary with the quality of the crude oil from which they were produced (363-390 g/mile). Heavier crudes require more upgrading and refining and produce less desirable by-products.

3. Tom McCann and Phil Magee of T.J. McCann & Associate Ltd., Calgary, "Crude Oil Greenhouse Gas Life Cycle Analysis Helps Assign Values For CO2 Emissions Trading," *Oil & Gas J.*, Feb. 22, 1999, pp. 38-44.

Full Life-Cycle GHG Emissions for FT & Petroleum Diesel Scenarios

For coal and biomass, the largest single source of emissions is the indirect liquefaction (FT conversion) facility (543 to 703 g CO_2-eq/mile), with GHG emissions even larger than those for end-use combustion. For pipeline natural gas, GTL emissions (121 g/mile) are lower than GHG emissions for end-use combustion. Carbon and oxygen must be removed from coal and biomass to convert them into a liquid. This step requires energy and consumes synthesis gas (H_2 and CO). The GTL process essentially extracts hydrogen from methane to produce liquid fuels. However, there is still a significant emissions penalty with GTL due to energy consumption during conversion. If the produced natural gas contains significant quantities of CO_2, emissions of GHG from conversion can be dramatically higher (212 vs. 121 g/mile, respectively). While combustion dominates total emissions for petroleum-based diesel, the other contributing sources are not insignificant. Conversion and refining emissions (74-143 g/mile), the second largest contributor, also vary with crude quality.

With improved fuel efficiency less fuel is consumed per mile and less fuel must be produced and transported. The net result of the adoption of next-generation advanced-diesel engine technology is an across the board 13% reduction in emissions per mile for all categories. This applies not only to the baseline IL #6 coal scenario, but to all the other scenarios listed above as well. In general, CP emissions from FT diesel combustion are lower than those from petroleum-derived diesel, making FT diesel an ideal alternative to petroleum-derived diesel in advanced engines.

While biomass conversion emissions are higher than those for coal (703 vs. 543-585 g CO_2-eq/mile); overall, the full-fuel chain GHG emissions for biomass-based FT fuels is very low (104 g/mile). Biomass is a renewable resource, and the carbon it contains is recycled between the atmosphere and the fuel, resulting in the fixation of 1011 g of atmospheric CO_2 in the biomass on a per mile basis. However, biomass cultivation and harvesting result in GHG emissions (42 g/mile), and biofuels should not be considered CO_2 emissions free.

The production of FT diesel from coal results in significantly higher total GHG emissions than those from petroleum-derived diesel (939-962 vs. 468-574 g CO_2-eq/mile). GTL technology can achieve GHG emissions levels between those for coal liquefaction and petroleum refining (562-652 g/mile), due to the higher hydrogen content of methane relative to petroleum (4 to 1 vs. ~2 to 1). In fact, the GHG emissions for FT diesel from natural gas are lower than the emissions for Venezuelan syncrude (562 vs. 574 g/mile) which requires severe processing to make it suitable as a feedstock for refining.

In some parts of the world, a significant amount of associated gas is flared, because there is no readily available market for this natural gas. When credit is taken for eliminating flaring, full fuel-chain emissions are cut drastically (from 643 to 65 g CO_2-eq/mile). The elimination of flaring and venting could under future regulations result in "carbon-credits" which could be sold in any market-based approach to reducing GHG emissions worldwide.

GHG Reductions Strategies
With the goal of identifying promising strategies for further study in mind, a preliminary examination was made of options for reducing GHG emissions from the production of FT derived fuels from coal. Material and energy balance models will be required to develop new conceptual designs for FT conversion processes employing these strategies and this will be the focus of future

work. The FT plant designs considered up to this point were developed in the early 1990s, when global warming was not yet considered a substantiated threat. As such, cost reduction was the major driver in the development of the conceptual designs, not GHG reduction or efficiency improvement.

Sensitivity Analysis: In order to help identify possible GHG reduction strategies for FT fuels production, a number of sensitivities were considered to the scenarios discussed above. These were particularly easy to estimate based on the detailed energy and material balances from the conceptual process designs. However, they only represent what may be possible, since they do not include any analysis (re-design) of the conceptual FT process they were based on. The sensitivities considered, in order of increasing GHG emissions reduction potential, are:

- Coalbed methane capture (maximum 2.3% reduction)
- Co-processing of coal and biomass (17%)
- Co-processing of coal and coalbed methane (25%)
- Co-production of fuels and power (32%)
- Sequestration of process CO_2 produced and vented during FT production (48%)
- Sequestration of process CO_2 and CO_2 from fuel combustion during FT production (55%)

Coalbed methane is released during coal mining and post-mining operations. While the magnitude of these releases is relatively small, the potency of methane as a GHG is quite high. Co-processing refers to the production of FT fuels from multiple feedstocks; for example, coal with methane and/or biomass. Since the latter have low GHG emissions relative to coal, co-processing has a moderating effect on the GHG emissions associated with FT fuels produced only from coal. Co-production refers to the production of multiple products from the indirect liquefaction plant; in this case, both fuels and power. Eliminating the recycle of off-gas produced in the FT conversion process, which can be used to produce electric power, reduces GHG emissions. Sequestration involves the collection, concentration, transportation and storage of CO_2 to reduce GHG emissions.

It is clear that many of the options discussed above will impose an energy and/or economic penalty on FT fuel production. For example, sequestration could require the compression of CO_2 for transportation and possibly for injection of CO_2 into any potential sink, and the production of nearly pure CO_2 from fuel combustion will require the increased production of high-purity oxygen at the FT plant. Increased energy requirements will result in increased CO_2 emissions from fuel combustion. It should be further acknowledged that economics might favor some of the options listed above with the least impact. For example, coalbed methane capture has an economic benefit in that coalbed methane can be sold as natural gas.

Based on potential economic, geographic and process synergies between the GHG reduction options listed above, estimates for three GHG reduction scenarios have been developed illustrating the incremental benefits of these options. These are:

- Co-processing of coal and biomass coupled with co-production of fuels and electric power and coalbed methane capture
- Co-processing of coal and biomass coupled with CO_2 sequestration and coalbed methane capture

- Co-processing of coal and coalbed methane from mined and unmined coal seams coupled with CO_2 sequestration in the unmined seams

The figure given on the following page illustrates the incremental benefits of combining GHG reduction strategies. The scenario involving coal and biomass co-processing coupled with sequestration shows the biggest GHG emissions reduction, 71% vs. 57% for biomass co-processing with co-production of power and 64% for coalbed methane co-processing with sequestration. To account for emissions penalties associated with implementing these strategies, rough estimates have been included for the efficiency of coalbed methane capture (80%), sequestration of process CO_2 (90%) and sequestration of CO_2 from combustion (80%).

All of the reduction scenarios achieve GHG emissions lower than those currently estimated for petroleum diesel fuel (286-442 vs. 468-574 g CO_2-eq/mile, respectively). However, it must be reiterated that *this analysis only identifies what may be possible*. Too much uncertainty exists in these estimates to consider any one of these scenarios better than another. Further detailed analysis will be needed to accurately quantify these future scenarios, and technology breakthroughs will be required in CO_2 sequestration, oxygen separation, and combustion technology to achieve these benefits. In addition, it must be kept in mind that petroleum production and refining would also benefit from similar strategies and technologies.

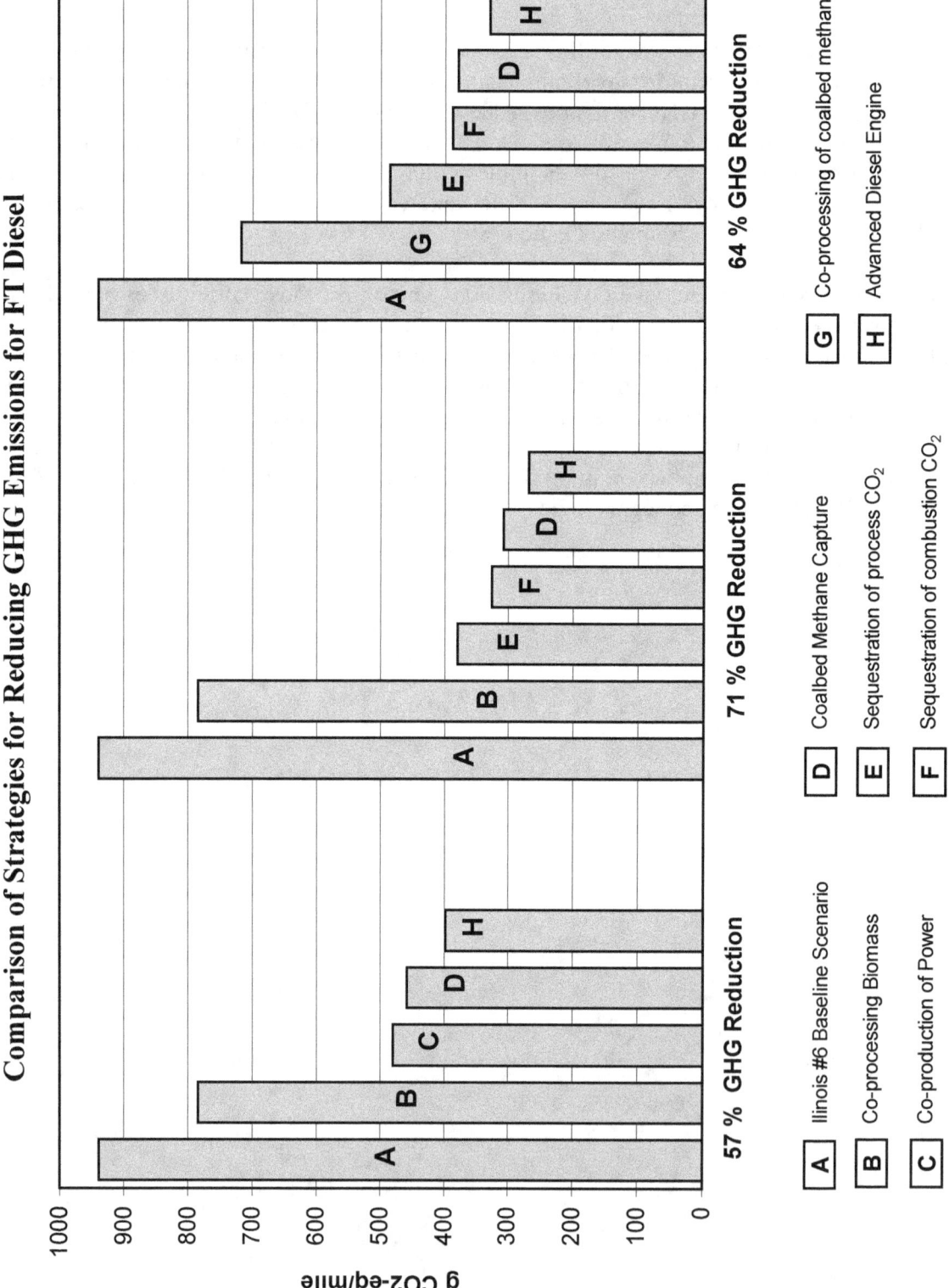

Comparison of Strategies for Reducing GHG Emissions for FT Diesel

Cost Impact: Many of the options considered here might be expensive to implement. Current estimates by Bechtel for the cost of indirect liquefaction correspond to a required selling price for the FT products of roughly $1.24 per gal (1998$s before taxes and marketing charges). However, there is reason to believe that rapid technology improvement in oxygen separation, coal gasification, and FT conversion could lower this price by as much as $0.20 per gal. This, coupled with the premium which FT diesel is likely to command, puts FT fuels in a near-competitive range with petroleum-derived gasoline and diesel.

Recent DOE estimates for the cost of sequestration technologies (other than forest sinks) are well over $100 per ton of carbon sequestered. The estimates for future technologies under development range anywhere from $5 to $100 per ton. The DOE carbon sequestration program has a goal of driving down the cost of sequestration to $10 per ton through aggressive technology development. While the CO_2 emissions from indirect coal liquefaction are high, the process has a significant advantage in that CO_2 can be removed from the process as a concentrated stream that could easily be sequestered. Based on these estimates then, the cost of CO_2 sequestration from indirect liquefaction is about $0.33 per gal based on $100 per ton and $0.02 per gal based on the DOE target of $10 per ton. The broad range of this potential added cost, and the possibility that it could wipe-out the significant cost reductions obtained over the last decade, *make it paramount that efforts to reduce the cost of FT conversion be continued.*

In the immediate future, only limited supplies of low-cost biomass are available for conversion. E^2S estimates the required selling price of FT fuels derived from biomass range anywhere from $2.00 to $2.30 per gal, depending on the source of the biomass. *Unless these costs can be reduced and the biomass resource base expanded, this option is likely to play only an incremental, albeit potentially important, role in GHG reduction strategies.*

The optimum coupling of all three strategies, sequestration, co-production, and co-processing, may be a very attractive GHG mitigation strategy to minimize both GHG emissions and their cost impact on indirect liquefaction. Thus, there is a pressing need to carefully examine in detail both the technology options for GHG emissions reduction and their cost impact on the FT product.

Conclusions & Recommendations

This analysis has identified and quantified significant sources of GHG emissions from the FT fuel chain. At present, GHG emissions from the FT fuel chain are greater than those from existing, petroleum-based fuel chain. Coal-based conversion is at a significant disadvantage relative to petroleum. Whereas, natural gas conversion is only moderately worse than the best petroleum scenarios and is better than the production and refining of heavy crude oils. In order for FT technology to be accepted in a world that is becoming more-and-more conscious of the effects of burning fossil fuels, it will be necessary to identify strategies and technologies for reducing these emissions. This study has been able to identify a number of possible approaches, including carbon sequestration, co-production of fuels and power, and co-processing of coal and biomass or coal and coalbed methane. Improvements in vehicle technology will also benefit the FT fuel chain by increasing fuel economy and, thus, reducing emissions per mile.

In order to evaluate the full potential of GHG reduction strategies for FT fuel production, all of the options considered here require better data and a more rigorous analysis beyond the scope of this preliminary analysis. Neither has a total view of the environmental benefits and deficiencies of FT fuels been realized in this study. A GHG emissions inventory has been completed, but only the first step has been taken toward developing a complete life-cycle inventory of all FT fuel chain impacts that affect the environment and human health and well being. Emissions of criteria pollutants have been identified for combustion sources along the fuel chain. Further work will be necessary to estimate emissions from vehicles fueled by FT diesel and gasoline and to expand this inventory to all categories of multimedia emissions.

TABLE OF CONTENTS

COMMON ACRONYMS

ANL	Argonne National Laboratory
ANS	Alaska North Slope
API	American Petroleum Institute
BCL	Battelle Columbus Laboratories
CP	Criteria Pollutants
DOE	U.S. Department Of Energy
EIA	DOE Energy Information Agency
EPA	U.S. Environmental Protection Agency
E^2S	Energy and Environmental Solutions, LLC
FT	Fischer-Tropsch
GHG	Greenhouse Gas
GTL	Gas-To-Liquid
GWP	Global Warming Potential
HAP	Hazardous Air Pollutants
HHV	Higher Heating Value
ILBD	Indirect Liquefaction Baseline Design
ISO	International Organization for Standardizations
LCA	Life Cycle Assessment
LCI	Life Cycle Inventory
LEBS	Low Emissions Boiler System
LHV	Lower Heating Value
LNG	Liquefied Natural Gas
LPG	Liquefied Petroleum Gas
MDEA	Methyl-Diethanol Amine
MF	Moisture Free
MTBE	Methyl Tert-Butyl Ether
NETL	National Energy Technology Laboratory
NG	Natural Gas
NGL	Natural Gas Liquids
NREL	National Renewable Energy Laboratory
NSPS	EPA New Source Performance Standards
PC	Pulverized Coal
POX	Partial Oxidation
PSA	Pressure Swing Absorption
SCOT	Shell Claus Offgas Treating
SETAC	Society of Environmental Toxicology and Chemistry
SUV	Sport Utility Vehicle
TAME	Tert-Amyl Methyl Ether
VOC	Volatile Organic Compounds
WGS	Water Gas Shift reaction

UNITS OF MEASURE

English units of measure have been used throughout the main body of this report. These are based on the units most commonly used to report specific data within the United States. For example, coal is commonly reported in "tons," crude oil in "barrels," gasoline in "gallons," etc. Appendix B gives the results from selected tables in standard Metric units. Given below are conversion factors for some units of measure frequently used in this report.

Mass: 1 Ton = 2,000 lb {pounds-mass} = 907.2 kg {kilograms} = 0.9072 Tonne {metric ton}

Energy: 1 Btu {British thermal unit} = 1,055.1 J {Joules} = 2.93×10^{-4} kWh {kilowatt-hours}

Distance: 1 mile = 5,280 ft {feet} = 1.6 km {kilometers}

Liquid Volume: 1 bbl {barrel} = 42 gal {gallons} = 5.615 ft^3 {cubic feet} = 159.0 l {liters} = 0.1590 m^3 {cubic meters}

Gas Volume: 1 scf {standard cubic foot @ $60^{\circ}F$ & 1 atm} = 26.8 Nl {Normal liters @ $0^{\circ}C$ & 1 atm}

Fuel Economy: 1 mpg {miles-per-gallon} = 0.4227 km/l {kilometers-per-liter}

Liquid Flowrate: 1 bpd {barrels-per-day} = 159.0 l/day {liters-per-day}

Temperature: $^{\circ}F$ {degree Fahrenheit} = $1.8 \times {}^{\circ}C$ {degree Celsius} + 32

API Gravity: $^{\circ}API$ = 141.5 / SpGr {specific gravity} - 131.5

English Prefixes: MM {million} = 1,000 M {thousand} = 1,000,000

Metric Prefixes: 1 T {tera} = 10^3 G {giga} = 10^6 M {mega} = 10^9 k {kilo} = 10^{12}

1. INTRODUCTION

The objective of this project was to develop a full life-cycle inventory (LCI) of greenhouse gas (GHG) emissions for synthetic fuels produced using the Fischer-Tropsch (FT) process. Where possible, emissions of criteria pollutants have also been compiled, and for the FT conversion process, a checklist of air toxics sources has been prepared. The LCI is based on a "cradle-to-grave" approach and includes data identification, collection and estimation of GHG emissions from upstream extraction/production, conversion/refining, transportation/distribution, and end-use combustion of FT fuels derived from three different feedstocks: coal, biomass and natural gas. *This inventory is the first step in a comprehensive strategy to identify, predict and reduce emissions from indirect liquefaction processes used for the production of alternative fuels.*

The scope of work included:
- Development of an inventory methodology for compiling and reporting GHG and other emissions for FT fuels and feedstocks [*Section 2*];
- Analysis of conceptual designs for FT conversion processes and estimation of significant process emissions [*Section 3*];
- Collection and evaluation of emissions data for all processes upstream [*Section 4*] and downstream [*Section 5*] of the FT conversion plant;
- Estimation of emissions from end-use fuel combustion and ancillary processes [*Section 6*];
- Compilation of emissions for the *full* FT-fuel life-cycle [*Section 7.1*];
- Analysis of baseline scenarios for the substitution of FT diesel fuel for petroleum-derived gasoline and diesel in SUVs [*Section 7.2*];
- Comparison of GHG emissions for FT diesel fuel with petroleum-derived diesel in SUVs [*Section 7.4*]; and
- Development of strategies and recommendations for reducing life-cycle GHG emissions from FT fuel production [*Sections 7.3 & 7.5*].

In this study, special emphasis was placed on estimating the projected emissions from FT process plants. Data collection activities did not involve field measurements of emissions. The FT plants considered are conceptual processes, which may be constructed in the near future. The material and energy balances used for the analysis are from designs developed for DOE by Nexant, Inc. (formerly a division of Bechtel Corporation) in the 1990s. Emissions from all processes upstream or downstream of the FT conversion plant where compiled from other sources, including a number of other life-cycle emissions inventory analyses conducted by ANL, EIA, EPA, NETL, and NREL.

The rigorous baseline scenarios analyzed in Section 7 are assembled by matching data compiled in Sections 3 through 6 for the different options for producing, transporting, delivering and utilizing FT fuels to the assumptions used for the various scenarios. The scenarios developed for reducing GHG emissions from FT fuel production are based on a sensitivity analysis of the baseline scenarios. These order-of-magnitude estimates for GHG reduction strategies indicate it is possible to significantly reduce GHG emissions from FT fuel production. *Further in-depth analysis will be needed to accurately quantify these GHG reduction scenarios, and technology breakthroughs will be required in CO_2 sequestration, oxygen separation, and combustion technology to achieve these benefits.*

2. INVENTORY METHODOLOGY

The objective of this project was to develop a full life-cycle inventory of greenhouse gas emissions for Fischer-Tropsch fuels. The life-cycle inventory is only the first component of a general procedure known as *life-cycle assessment*. Life Cycle Assessment (LCA) is an analytical approach for qualifying and quantifying the environmental impacts of all processes used in the conversion of raw materials into a final product. LCA dates back to the late 1960s/early 1970s and has also been described as *full fuel-cycle* analysis, *ecobalancing* or *cradle-to-grave* analysis. What is conveyed by these names is that LCA attempts to quantify all significant impacts which arise from raw materials acquisition, manufacturing, transportation, use/reuse/maintenance, and recycle/disposal of a given product or service. It is increasingly becoming understood within policy circles that from a socio-economic perspective, any comparison of the environmental impacts from different products or services may be meaningless, or worse misleading, if only "across-the-fence" plant emissions are considered and all other impacts are ignored. LCA attempts to account for all consequences.

Broadly, LCA can be broken down into three distinct activities: *inventory analysis*, *impact assessment* and *improvement analysis*. Life-Cycle Inventory (LCI) Analysis catalogs and quantifies all materials and energy used and the environmental releases arising from all stages of the life of a product, from raw material acquisition to ultimate disposal. Life-Cycle Impact Assessment evaluates actual and potential environmental and human health consequences and resource depletion from (*that is, sustainability of*) all activities identified in the inventory phase. Life-Cycle Improvement Analysis aims at reducing any risks identified in the impact assessment, possibly by modifying stages in the product life cycle.

Prior to beginning an LCA, careful consideration must be given to the scope of the study. *Scope Definition* includes clearly identifying the purpose of the study (*What will it be used for?*) and identification of all assumptions to be used in, or restrictions to be placed upon, the assessment. Items to be considered include the selection of system boundaries; availability, quality and level of aggregation of data; classification and characterization of emissions; and the allocation of impacts to multiple products.

Within the U.S., the Society of Environmental Toxicology and Chemistry (SETAC) has been working to establish a standard framework for conducting LCA [1-4]. The International Organization for Standardization (ISO) has also developed a protocol for LCA as part of its ISO 14000 environmental management standards [5][4]. The framework used here has been adapted from these standards and protocols to reflect the needs of the National Energy Technology Laboratory's research programs. NETL is not a regulatory organization concerned with labeling products and procedures for the consumer. *This assessment is focused on making relative comparisons of existing and future technologies for producing transportation fuels, with the goal of improving these technologies through applied R&D.*

---

4. Information on the ISO 14000 Environmental Standards (EMS) can be accessed via *www. iso.ch.com* or *www.iso14000.com*.

The analysis reported here is a *full* LCI in the sense that the emissions being cataloged are tracked from cradle to grave. It includes emissions from *upstream extraction/production, conversion/refining, transportation/distribution,* and *end-use fuel combustion.* However, the LCI is not a *complete* inventory since only greenhouse gases and criteria pollutants were quantitatively considered, and air toxics are only covered qualitatively (that is, only a list of the compounds that must be reported to the EPA has been prepared). It should not be confused with or substituted for a complete LCA, since it does not meet the SETAC criteria of being multi-media in perspective, nor does it include rigorous impact assessment or improvement analysis. This said, the analysis does consider two important elements of impact assessment, classification and characterization of the GHG emissions cataloged. Neither has improvement analysis been completely ignored. During this inventory, several approaches became obvious for reducing GHG emissions from the FT fuel chain. Order-of-magnitude estimates for these promising reduction approaches are included in this report.

2.1   System Boundaries

Figure 1 shows the *fuel chain* associated with the production of liquid fuels based on the Fischer-Tropsch process. A two-tiered approach has been taken for the collection and organization of emissions inventory data for the fuel chain. All material and energy use and environmental releases along the fuel chain are classified as either *primary* or *ancillary*. This *streamlining* procedure has been used to simplify this analysis while still identifying and quantifying all significant impacts. Primary emissions result from the actual operation of the process steps making up the major systems identified in Figure 1. They are designated primary because they result from the processing of the primary resources, which in the cases considered here are coal, biomass and natural gas. Primary emissions occur on the direct path from cradle to grave. The designation primary is not intended to imply that these flows are always significant in relation to the entire life cycle. For example, CO_2 emissions from transport of gasoline between storage-terminal tankage and service (re-fueling) station are usually not significant relative to the entire fuel chain. However, they have been included for completeness in this LCI. Ancillary material and energy use and environmental release are *aggregated* data for all activities associated with the external flows into the major systems of the FT fuel chain (that is, the ancillary feedstocks). Ancillary emissions are included in the inventory unless otherwise noted and, in some cases, may be significant.

As indicated by Figure 1, the steps in converting the primary resource into the final product, transportation miles, are the same regardless of the feedstock: coal, biomass or natural gas. The first step is mining for coal, cultivation and harvesting for biomass, and oil and gas production for natural gas. The second step is conversion. For FT-based conversion to fuels, this step involves gasification of coal or biomass and partial oxidation/reforming for natural gas. The resulting syngas (synthesis gas, a mixture containing H_2 and CO) is then converted via Fischer-Tropsch synthesis into liquid hydrocarbons suitable for the manufacture of fuels and chemicals. This conversion step is often referred to as indirect liquefaction for coal and biomass and gas-to-liquid conversion for natural gas.

Primary Resources

Coal   Biomass   Natural Gas

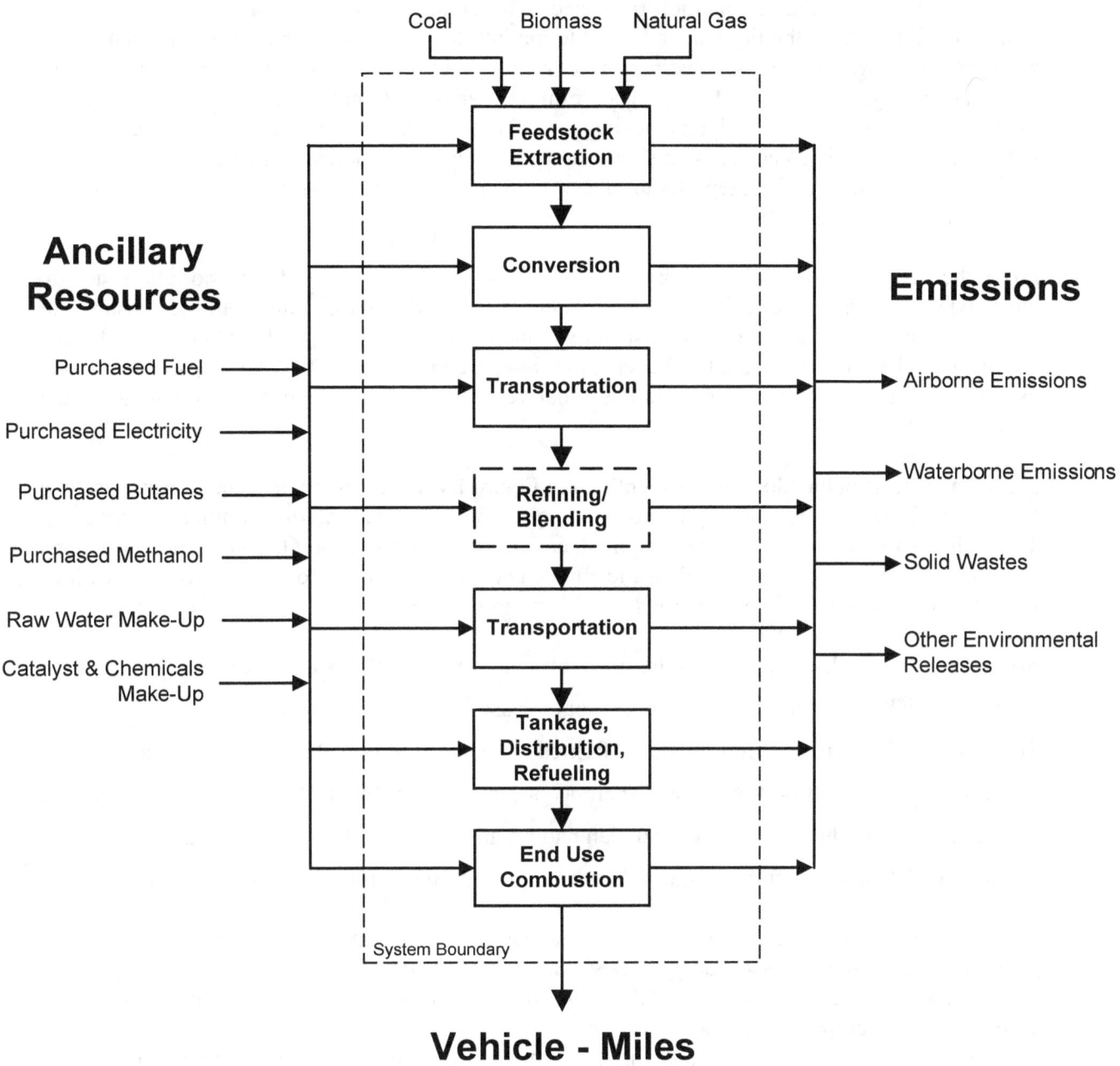

Ancillary Resources

Purchased Fuel

Purchased Electricity

Purchased Butanes

Purchased Methanol

Raw Water Make-Up

Catalyst & Chemicals Make-Up

Emissions

Airborne Emissions

Waterborne Emissions

Solid Wastes

Other Environmental Releases

Feedstock Extraction

Conversion

Transportation

Refining/ Blending

Transportation

Tankage, Distribution, Refueling

End Use Combustion

System Boundary

Vehicle - Miles

Figure 1. FT Fuel Chain

5

It is assumed in all the scenarios considered here (with the exception of the scenario based on pipeline gas) that the conversion step occurs in close proximity to feedstock extraction and remote from the end-use markets for the fuels produced. Thus, one step involves the transportation of the synthetic FT fuel from the liquefaction plant to market. In reality, a number of intermediate steps occur along the way, possibly including further refining of the raw FT fuel into specification fuels (e.g. gasoline, jet and diesel fuel). The refining step might include processes as severe as hydrocracking and/or fluid catalytic cracking or as simple as blending with refined petroleum fuels. In Figure 1, the refining step has been shown as a dashed block to indicate that it may or may not be distinct from the conversion step. Examples of both situations are found in the FT design options considered.

From the refinery, the specification fuels are transported in a second transportation step to intermediate storage and distribution centers (tank farms) for final distribution to the consumer at service/re-fueling stations. Tankage, distribution and refueling are lumped together as a sixth step in Figure 1. The final step in the FT fuel chain is end-use combustion. This LCI focuses on the final use of these fuels for transportation, in particular vehicles employing conventional and advanced diesel engines.

Particular aspects of the blocks/steps identified in Figure 1 will depend on both the starting resource and the final fuel product and application (e.g., gasoline and diesel internal combustion engines). They will also vary based on the geographic locations of the resource and the fuel market. Among other things, these locations establish the routes and methods required to transport the various intermediates. The fuel chain scenarios considered in this analysis are:

Scenario 1: FT production from southern Illinois coal for use in the Chicago area

Scenario 2: FT production from Wyoming coal for use in the Chicago area

Scenario 3: FT production from biomass, farmed in southern Illinois, for use in the Chicago area

Scenario 4: FT production from pipeline natural gas, in southern Illinois, for use in the Chicago area

Scenario 5: FT production from Venezuelan natural gas for use in the Chicago area

Scenario 6: FT production from Alaska North Slope natural gas for use in the San Francisco area

These baseline scenarios are assembled from the various FT design, feedstock, transportation and distribution and end-use options analyzed. Sensitivities were considered for some of these scenarios to examine the effect on life-cycle GHG emissions of sequestering CO_2 produced in the FT conversion step, co-producing fuels and power, co-feeding coal and biomass, co-feeding coal and coalbed methane, capturing coalbed methane, and mitigating natural gas venting and flaring. Further in-depth analysis will be required to accurately quantify the more promising of these strategies for reducing GHG emissions. More detailed descriptions of the various blocks shown in Figure 1 are given in Sections 2 through 6 of this report.

2.2   Classification & Characterization

Classification is the process of assigning an inventory result to an appropriate *impact* or *stressor category* and *characterization* involves converting individual results for a category into a *category index* or *equivalency factor*, possibly based on a conceptual environmental mechanism.

The impact categories of primary interest for this study are greenhouse gases (GHG), criteria pollutants (CP), and air toxics. The greenhouse gases considered are: CO_2 (carbon dioxide) from fossil-fuel combustion along the life cycle and venting from natural gas production; CH_4 (methane) from fugitive plant and pipeline emissions, incomplete combustion or incineration (gas flaring), and coalbed methane releases; and N_2O (nitrous oxide) from fuel combustion and the cultivation of biomass feedstocks. Other gases such as chlorofluorcarbons, while extremely potent greenhouse gases, are not used or released in significant quantities from the processes of interest to warrant inclusion in this inventory.

The current interest in greenhouse gases is driven by concerns over the effect that a buildup of these gases in the atmosphere may have on the Earth's climate. The "greenhouse-effect" is proven. The greenhouse gases mentioned above (and others) prevent the sun's radiant energy from being entirely re-radiated back into space as infrared radiation, by absorbing some of this radiation. Human activities in the last two centuries (since the onset of the industrial revolution) have resulted in increasing concentrations of certain greenhouse gases in the atmosphere, thus possibly trapping more solar energy and raising the global average temperature. The effects of such an increase in temperature on the planet can only be predicted by computer simulation. Examining the geological record from previous cycles of planet-wide warming and cooling can give some clues at to what may happen.

While predicting climate change is tremendously complex and many phenomena are still poorly or not understood, efforts have begun worldwide to decrease the rate of increase of GHG emissions. Each greenhouse gas absorbs radiation in a particular set of wavelengths in the spectrum and therefore, individual gases can have very different heat-trapping effects. In order to quantify the heat-trapping effects, assess progress and establish targets, emissions of individual greenhouse gases are characterized into a single metric called the *Global Warming Potential* (GWP). The purpose of the GWP concept is to account for the <u>relative</u> impacts on global warming of various gases compared to carbon dioxide on a weight basis (kg-per-kg). Carbon dioxide, which is the greenhouse gas produced in the largest quantity by the burning of fossil fuels and the least effective greenhouse gas in trapping the Earth's radiant heat, is used as a reference and assigned a GWP of 1.0. The value of a gas's GWP is also a function of the "atmospheric lifetime" or the period of time it would take for natural processes (decomposition or absorption into the ocean or ground) to remove a unit of emissions from the atmosphere. For example, gases such as chloroflurocarbons have lifetimes in hundreds of years whereas carbon monoxide has a lifetime measured in hours or days. Table 1 contains the GWPs recommended by the *Intergovernmental Panel on Climate Change* for the three greenhouse gases of interest in this study: CO_2, CH_4 and N_2O, using three time horizons 20, 100 and 500 years. For example, although methane's atmospheric lifetime is 12 years, its GWP for a 100 year time horizon is still 21 times greater than carbon dioxide; or 10 kg of CH_4 will have a heat-trapping effect equivalent to 210 kg of CO_2 in 100 years. The GWP values for the 100-year time horizon, referred to as *Greenhouse Gas Equivalency Factors*, are used in this study; though, the results could easily be updated to consider other horizons. Examples of these calculations are given in Appendix A.

Table 1: Global Warming Potentials for Selected Gases*
(kg of CO_2 per kg of Gas)

Gas	Lifetime (years)	Direct Effect over Time Horizons of:		
		20 Years	100 Years	500 Years
Carbon Dioxide (CO_2)	Variable	1	1	1
Methane (CH_4)	12 ± 3	56	21	7
Nitrous Oxide (N_2O)	120	280	310	170

*as reported in [6]

Data were also compiled, where possible, for airborne emissions of CO (carbon monoxide), NOx (Nitrogen Oxides), SOx (Sulfur Oxides), VOC (Volatile Organic Compounds), and PM (Particulate Matter). The U.S. EPA classifies these substances as criteria pollutants (CP). At the level of detail of this study, it was not possible to speciate VOCs or further sub-classify PM. There is overlap between the GHG and CP categories. Methane is both a greenhouse gas and a VOC. Other criteria pollutants are believed to participate in global warming; however, the mechanism is not well understood, and they have not been included in the GHG impact category. The only source of CP considered here is combustion. SOx emissions (calculated as SO_2) result from oxidation of sulfur present in fuel. NOx emissions (calculated as NO_2) are the result of both the oxidation of nitrogen in fuel and thermal conversion at high temperatures of N_2 present in combustion air. Emissions of CO, VOC and PM result from incomplete combustion of fuels. PM emissions also result from ash liberated from the fuel during combustion. CP emissions from all combustion sources along the FT fuel chain up to the point of sale of the fuel products have been included in the inventory. CP emissions from end-use combustion of FT fuel are more difficult to analyze, since cars and trucks normally operate under variable loads. Further work will be needed for their incorporation into the LCI.

A checklist was also prepared of compounds used or produced in FT conversion processes, which have been identified by the U.S. EPA as air toxics and hazardous air pollutants (HAPs). Emissions of these substances must be reported to the EPA annually. While these compounds may be released as airborne emissions, no effort has been made to estimate what their emissions might be for the conceptual FT processes studied. Neither have checklists of this kind been developed for the processes upstream and downstream of the FT plant.

No attempt has been made here to characterize individual airborne pollutants as smog precursors, for acidification potential, etc.; or have the results of the inventory been normalized (*normalization* involves dividing an indicator/index by some reference value, commonly the total loading for the given category) or been subject to any valuation (*valuation* involves formalized ranking or weighting to aggregate indicators/indices across multiple categories into a final score). These refinements were considered to be outside the scope of this analysis.

2.3   Impact Allocation

It is standard practice for life-cycle inventory analysis to *allocate* impacts, such as emissions, between the product and various by-products that are generated during the life cycle of the product, though there is some debate on how to actually do this. This procedure, however it might be implemented, is likely to be adequate, if the by-product production rates are relatively small, but this is generally not the case for the energy and fuel systems considered here. Existing petroleum refineries have multiple products, sold for a variety of applications, and future energy systems now being considered may produce electric power in addition to liquid fuels. FT conversion processes also result in a multitude of products, some of which are not used in transportation.

Careful consideration was given to how emissions should be allocated between the various FT fuel products. *For this study, it was decided to allocate emissions from conversion, refining, and all other upstream operations between the LPG, gasoline and distillate fuel products based on the ratio of the energy content (LHV) of the specific fuel relative to the total product.* It is unlikely that more complicated procedures would result in substantially different results, since the energy densities of these liquid fuels are similar. However, this procedure was not considered appropriate when electric power was produced as a major by-product of FT production, since in some sense, power can be considered an end use for all FT fuels produced. To compensate for this, *emissions are allocated to power based on the energy content of the fuel used in the electrical conversion device* (gas or steam turbine); that is, the energy content of the electrical power is divided by turbine efficiency when determining the share of emissions to be allocated to this power. This is similar to the procedure used when calculating the *thermal efficiency* of co-generation (power & steam) processes.

In order to compare the inventory results from the various scenarios considered here, it is necessary to select a *functional unit* to use when reporting results. The functional unit is the production amount that represents the basis of the analysis. This might be gallons of total LPG, gasoline and distillate fuel produced; standard cubic feet of syngas converted; or total energy contained in the products produced. However, it can just as readily be miles of transportation provided or kWh's of electricity delivered. These are services as much as they are tangible products. For the case study reported in Section 7, substitution of FT diesel fuel in diesel-powered SUVs, a *per-vehicle-mile driven* basis was used. Fuel economies in miles-per-gallon (mpg) were used to convert emissions from a per-gallon to a per-mile basis. Since inventory results will change based on the fuel economy used for this conversion, the comparison is specific to SUV conversion from conventional gasoline engines to conventional and advanced diesel engines and is not applicable to passenger cars, heavy-duty trucks, etc. For heavy construction equipment, a better functional unit would be brake horsepower-hr, since this is a measure of the total work being performed.

In general, common English units have been used in the main body of this report. Appendix B gives the results from selected tables in Metric units. The units used to report emissions in the main body of this report are g/ton (MF, moisture free) for coal and biomass production, g/Mscf for natural gas production, g/bbl for FT fuel production and ancillary feedstocks, g/gal for FT fuel transportation,

9

and g/MM Btu for ancillary fuel consumption. For the full inventory reported in Section 7, both g/gal of FT fuel delivered and g/mile driven are reported.

2.4 Inventory Data Issues

Inventory analysis is primarily data driven and results in a database, which is accessed and used in the other phases of LCA. Ideally, one would want these inventory data to be as complete and as accurate as possible, regardless of the scope of any assessment to be performed using these data. This, however, is not often possible, and limitations of the data do impact scope, to varying degrees, for any particular analysis. Data for the inventory can come from measurements done on actual systems or may be the output obtained from process simulation and modeling. Measured data are preferable, but not always available. Both types of data are used here; however, *since the fuel technologies of interest to this study are not widely commercialized* (if at all), *there is a heavy dependence on modeling results and estimated emissions.*

Data collection activities did not involve actual field measurements of emissions. Input data for the inventory were collected from available literature sources and through direct contact with experts in various fields, such as oil tanker transportation, trucking and coal mining. In many instances, the emissions have been estimated either directly by the authors or indirectly by the suppliers of this information. *Special emphasis was placed on estimating the projected emissions from FT process plants.* Emissions from all processes upstream or downstream of the FT conversion plant where compiled from other sources, including a number of other life-cycle emissions inventory analyses conducted by ANL, EIA, EPA, NETL, and NREL. Efforts were made to validate emissions data by comparing data from multiple sources; nevertheless, many inconsistencies remain, and some data are controversial. Data that are missing or considered uncertain have been marked in the appropriate tables as 'na' (not available).

In general, impacts of upstream processes become less significant in the analysis the further one proceeds away from the process of interest (both temporally and spatially), and a trade off becomes apparent between time and effort spent and detail and accuracy of the final inventory. Since the FT processes of interest are still conceptual, little accuracy and relevance are gained by including emissions associated with the manufacture and construction of capital equipment. The minimum useful life of a FT facility would be 20 years or more. However, when considering end-use of the FT fuel, the situation is more complex. The useful life of transportation vehicles, in particular personal automobiles and SUVs, is measured in terms of a few years instead of tens of years, and vehicle replacement and maintenance (such as replacement of tires and engine oil) will impact life-cycle emissions [20]. These effects have been neglected with the caveat that the comparisons made here are between conventional vehicles with similar life expectancies and maintenance requirements and not between radically different vehicle systems (e.g., electric or hydrogen powered vehicles).

In regard to emissions from ancillary resources, the LCI analysis has also been simplified. Upstream emissions from ancillary feeds to the FT fuel chain have either been estimated from available data or, in some cases, completely ignored based on the relative magnitude of the in-flow to the FT fuel chain. Section 6 - *Fuel Combustion, Efficiencies, & Ancillary Emissions* gives explicit information on which emissions have been included for what resources.

10

Special note must be made relative to the effects of scale. Resources consumed, energy used, and emissions are all functions of the size of the plant being considered, with larger facilities, in general, being more efficient. The FT process designs used here are for plants with nominal capacities of 50,000 bpd of FT product with the exception of the biomass-based conversion plant, which produces only about 1,200 bpd. Care should be exercised when comparing results from cases with widely varying throughputs.

Since only greenhouse gases and criteria pollutants are considered in this study, it has been relatively easy to perform inventory collection and analysis using simple spreadsheet models versus using specialized software packages. Estimating procedures along with sample calculations appear in Appendix A.

3.   FISCHER-TROPSCH PROCESS

The Fischer-Tropsch (FT) synthesis was discovered in the 1920s by the German chemists F. Fischer and H. Tropsch. It was briefly used by Germany before and during World War II to produce fuels, and has generated varying levels of interest worldwide since that time. Today, it is used commercially to produce transportation fuels and chemicals at several sites in South Africa, both from coal and natural gas, and at a single site in Malaysia from natural gas. However, there is considerable interest in this technology for the conversion of stranded natural gas reserves into an easily transportable, liquid product.

The FT synthesis involves the catalytic reaction of H_2 (hydrogen) and CO (carbon monoxide) to form hydrocarbon chains of various lengths (CH_4, C_2H_6, C_3H_8,...). A major by-product from the reaction is water. The FT synthesis reaction can be written as:

$$(n/2 + m)\, H_2 \quad + \quad m\, CO \quad \rightarrow \quad C_mH_n \quad + \quad m\, H_2O$$

where m is the average chain length of the hydrocarbons formed, and n equals $2m+2$, if only paraffins are formed, and $2m$, if only olefins are formed. Temperature is one of the main variables affecting the value of m. For iron catalysts, the value of n is intermediate, and a mixture of n-paraffins and n-olefins results with small quantities of n-alcohols also synthesized. Iron has water-gas shift (WGS) activity, which converts much of the water of reaction into CO_2, (carbon dioxide), generating additional H_2. The WGS reactions is:

$$CO \quad + \quad H_2O \quad \leftrightarrow \quad CO_2 \quad + \quad H_2$$

Therefore, synthesis gases with a wide range of H_2 to CO ratios may be used as feed to the FT synthesis, and the WGS reaction can be used to adjust the H_2 to CO ratio to match requirements for hydrocarbon synthesis. Syngas can be produced from coal and biomass by means of gasification. In gasification, oxygen is reacted with the feedstock under conditions which result in partial oxidation (POX) of the feed to form H_2, CO, CO_2, H_2O, CH_4, and small quantities of other hydrocarbon gases. Impurities in coal and biomass also result in the formation of H_2S, NH_3, HCl, and other trace substances that must be removed prior to the FT synthesis. The H_2 to CO ratio for syngas from the coal and biomass gasifiers considered in this study is less than 0.7, and steam is injected into the FT reactor to promote the production of additional H_2 via the WGS reaction.

Synthesis gas derived from natural gas typically has a much higher H_2 to CO ratio than that produced by gasification of coal and biomass, a result of the higher hydrogen content of CH_4 (methane), the primary constituent of natural gas. Natural gas is converted to syngas either by partial oxidation, steam reforming, or a combination of both called autothermal reforming. The exothermic POX reaction of methane is:

$$2\, CH_4 \quad + \quad O_2 \quad \rightarrow \quad 2\, CO \quad + \quad 4\, H_2$$

13

In the endothermic reforming reaction, oxygen for syngas production is supplied by H_2O (steam) instead of by O_2 from air separation. This reaction is:

$$CH_4 \ + \ H_2O \ \rightarrow \ CO \ + \ 3\,H_2$$

Cobalt catalysts are typically used to convert this high H_2 to CO ratio (~2:1 for POX and ~3:1 for reforming) syngas to hydrocarbons. Cobalt catalysts do not have WGS activity, and water is the primary by-product of the FT synthesis. Paraffins are the dominant hydrocarbon products with only lesser quantities of olefins and alcohols being formed. The H_2 to CO ratio required for the FT synthesis reaction then is $(2m+1)/m$ or $2+1/m$. The H_2 to CO ratio of syngas produced from natural gas can be adjusted to meet this requirement either by externally shifting the syngas or using a combination of POX and steam reforming. If the later is accomplished within a thermally integrated reactor, it is known as autothermal reforming.

The biomass design considered in this study employs an indirectly heated gasification process. The biomass is gasified with steam (reformed) in a fluidized bed of inert sand particles. During this process char is formed. A slipstream of char and sand is removed from the reforming bed and fed to a second fluidized bed where the char is combusted with air. The hot clean sand is then re-circulated to the first bed and provides the necessary heat for the reforming reactions.

The FT reactor considered in this study is a slurry bubble-column reactor. In the slurry bubble column, syngas is bubbled through a suspension of fine catalyst particles. The FT synthesis products distribute between the vapor and liquid phases within the reactor. The lighter hydrocarbons are carried overhead with unreacted syngas, and the heavier components form the molten-wax phase within which the catalyst is suspended. The slurry bubble column is not the only reactor system that can be used for the FT synthesis; fixed catalyst bed and fluid bed systems are used commercially.

The liquid hydrocarbon products from the FT synthesis are of high quality, having negligible sulfur, nitrogen or aromatic impurities and high hydrogen content. They can be transformed into clean-burning transportation fuels by a variety of refining routes. The lighter (lower-boiling) liquid is referred to as naphtha and is a feedstock to a number of processes for producing gasoline-blending components. The heavier (higher boiling) liquid is referred to as distillate. It is generally of sufficient quality to be used directly as a premium diesel fuel, but also may be blended with other distillate fuels to improve their overall quality. The heaviest hydrocarbons formed in the synthesis are a solid wax at ambient conditions and must be cracked to produce liquid products. The lighter C1-C4 gaseous hydrocarbons produced by the synthesis can be recycled back to the syngas generation step or burned in a fired-heater to fulfill plant process heating requirements or in a gas turbine to produce electricity for plant utility requirements (or for sale). C3-C4 hydrocarbons may also be recovered and sold as LPG (Liquefied Petroleum Gas) or converted to high-value gasoline blending components.

3.1 Indirect Liquefaction Baseline Designs

In 1991, Bechtel, along with AMOCO as a major subcontractor, was contracted by the DOE (DE-AC22-91PC90027) to develop conceptual designs, economics and process simulation models for indirect liquefaction based on advanced gasification and Fischer-Tropsch technology. The original focus of these projects was coal liquefaction using two grades of coal, bituminous Illinois No. 6 and subbituminous Powder River Basin. Several design options were also included. The study was later expanded several times to include other design options, primarily related to the upgrading of the FT reactor liquids, and also to consider natural gas based FT synthesis, so-called Gas-To-Liquid (GTL) technology. A final report on this project was issued in April 1998 [7].

Bechtel and its subsidiary, Nexant, Inc., were also contracted to perform other related projects for DOE (DE-AC22-93PC91029). One involved indirect liquefaction of biomass to produce FT liquids and another development of an updated and improved GTL design. Topical reports for these projects were issued in May 1998 [8], and December 2000 (draft) [9].

The Indirect Liquefaction Baseline Design (ILBD) cases developed by Bechtel/AMOCO form the basis for the emissions estimates developed in this report. A description of the design options follows:

Option 1 – Illinois No. 6 Coal with Conventional Product Upgrading (maximum distillate production) [Case 1 from Bechtel report, 7]

Option 2 – Illinois No.6 Coal with Alternate ZSM-5 Product Upgrading (increased gasoline production) [Case 2 from Bechtel report, 7]

Option 3 – Illinois No. 6 Coal with Conventional Product Upgrading (maximum gasoline & chemicals production) [Case 5 from Bechtel report, 7]

Option 4 – Wyoming Powder River Basin Coal with Conventional Product Upgrading (maximum distillate production) [Case 3 from Bechtel report, 7]

Option 5 – Biomass with Conventional Product Upgrading and Once-Through Power Generation [8]

Option 6 – Pipeline Natural Gas with Conventional Product Upgrading (1990 technology - maximum distillate production) [Case 7 from Bechtel report, 7]

Option 7 – Associated Natural Gas with Conventional Product Upgrading (2000 technology - minimum upgrading) [9]

Option 8 – Associated Natural Gas with Conventional Product Upgrading and Once-Through Power Generation (2000 technology - minimum upgrading) [9]

The eight design options listed above differ in a number of significant ways. Five different feedstocks are represented: two coals, Illinois No. 6 bituminous coal (Options 1-3) and Wyoming subbituminous coal (Option 4); biomass, maplewood chips (Option 5); and two natural gas compositions, pipeline specification gas (Option 6) and associated gas from oil production (Options 7 & 8). The coal and biomass based designs employ iron FT catalyst; whereas, the natural gas based designs use cobalt. The Shell gasification process was used in the coal designs, the BCL gasification process in the biomass design, a combination of POX and steam reforming in the pipeline gas design, and autothermal reforming in the associated gas designs. Autothermal reforming is also used in all the coal designs to convert light hydrocarbons (CH_4, C_2H_4, and C_2H_6) back into syngas for recycle to the FT reactor.

The eight design options also differ in the extent and complexity of upgrading used to convert the raw FT reactor liquids to fungible products. Options 1, 4, 5 and 6 all employ conventional refining technology which includes extensive hydroprocessing of the raw liquids. Hydrocracking is used for the conversion of wax to naphtha and distillate. These designs maximize the amount of distillate fuel produced. Option 3 also employs conventional refining technology; however, fluidized-bed catalytic cracking is used for wax conversion. This increases the yield of gasoline relative to distillate fuel and produces propylene for chemical sales. In Option 2, the Mobil ZSM-5 process is employed to directly convert the vapor stream leaving the FT reactor into a premium gasoline blending component. This also increases the yield of gasoline relative to distillate. Options 7 and 8 contain minimal upgrading of the raw FT liquid. Only, hydrocracking is used to convert the wax into additional naphtha and distillate. No other refining is used to upgrade the products. These two designs are more indicative of situations that might arise where the size of the FT plant does not warrant the addition of capital intensive refinery processing, or of locations where the FT product will be shipped to remote markets. Options 5 and 8 also co-produce electric power, which simplifies the overall plant design. Plant location plays a significant factor in all of the designs.

Improvements in process technology are also represented in the design options. The natural gas Options 6, 7 and 8 differ in degree of technology advancement considered. Option 6 is a snapshot of gas-to-liquid technology *circa.*1990. Options 7 and 8 are representative of the state-of-the art in autothermal reforming, FT slurry-bubble column design, cobalt catalyst and hydrocracking technology *circa.* 2000. The remaining designs also represent "older" technology, and it is likely that updated designs would include significant changes to the gasification and FT synthesis processes.

A summary of the design conditions for the eight options considered is given in Table 2.

Table 2: Indirect Liquefaction Baseline Design Data*

Design	Option 1	Option 2	Option 3	Option 4	Option 5	Option 6	Option 7	Option 8
Feedstock	IL #6	IL #6	IL #6	Wyo. Coal	Biomass	Pipeline Gas	Assoc. Gas	Assoc. Gas
Upgrading	Maximum Distillate	Increased Gasoline	Maximum Gaso. & Chem.	Maximum Distillate	Fuels & Power	Maximum Distillate	Minimum Upgrading	Min. Upgrading & Power
Raw Materials								
Coal/Biomass/NG (MF ton/day)	18575	18575	18575	19790	2205	8949	13781	13781
Natural Gas (Mscf/day)						412	507	507
Catalysts & Chemicals (ton/day)	342	384	na	394	na	2.92	na	na
Products (bbl/day)								
Methanol			-2303					
Propylene			5060					
LPG	1922	2623	1573	1907	0	1704	0	0
Butanes	-3110	998	-5204	-3101	0	-340	0	0
Gasoline/Naphtha	23943	31255	39722	23756	382	17027	15400	12100
Distillates	24686	15858	9764	24466	775	26211	33800	26700
Products (ton/day)								
Methanol			-321					
Propylene			460					
LPG	171	233	140	169	0	151	0	0
Butanes	-317	102	-531	-316	0	-35	0	0
Gasoline/Naphtha	3021	3904	4988	2997	49	2153	1853	1456
Distillates	3343	2162	1302	3313	105	3542	4548	3586
By-Products								
Slag (MF ton/day)	2244	2244	2244	1747	230			
Sulfur (ton/day)	560	560	560	108				
CO$_2$ Removal (ton/day)	28444	28414	28463	28325		3270	5114	
CO$_2$ Carrier Gas (ton/day)	-3715	-3715	-3715	-3958				
S-Plant Flue Gas (ton/day)	1086	1086	1086	348				
Utilities Consumed								
Electric Power (MW)	54.3	53	58	88	-86	-25	0	-372
Raw Water (MM gal/day)	14	14	16	10	2	21	6	4

*Negative products/byproducts are consumed, negative utilities are produced; data from [7-9].

17

3.2 Process Flowsheet Descriptions

While the design options described in the preceding section differ in details, they can be broken down into four main plant areas: the Syngas Generation Area, which varies based on the nature of the feedstock; the FT Conversion Area, which varies based on the nature of the catalyst; the FT Product Upgrading Area, which varies based on the nature of the final products desired; and Offsite supporting systems. The following sections describe the different process flowsheets developed by Bechtel. The reader not interested in the details of the designs may wish to skip directly to Section 3.2.4

3.2.1 Coal Based Designs

The designs considered in Options 1-4 are all variations on the block flow diagram shown in Figure 2. A breakdown of the various process plants appearing in Option 1 - Illinois No. 6 Coal with Conventional Product Upgrading (maximum distillate production) is given below:

Syngas Generation Area

Coal Receiving & Storage (not shown in Figure 2) - Receives washed coal from mine-mouth coal washing plant, stores the coal in piles, reclaims the coal from storage, and delivers coal to the coal preparation plant.

Coal Preparation - Dries and grinds the coal for use in coal gasifiers.

Air Separation - Provides high-purity (99.5%) oxygen, using cryogenic air separation, for gasification and autothermal reforming of recycle gas.

Gasification - Pressurizes and feeds prepared coal to Shell gasifiers and gasifies coal; includes gas quench, high-temperature gas cooling, slag handling, fly-slag removal and handling, and solid waste handling. CO_2 is used as the carrier gas for the feed coal.

Syngas treatment includes the following three plants:

Syngas Wet Scrubbing - Removes trace amounts of fine particles and humidifies the syngas.

COS Hydrolysis & Gas Cooling - Converts COS to H_2S, HCN to NH_3, and cools the syngas.

Acid Gas Removal - Selectively removes H_2S from the syngas using amine solvent; solvent is regenerated and H_2S-rich gas sent to sulfur recovery.

Sulfur Guard Bed - Removes trace amounts of sulfur compounds, including H_2S, COS and CS_2, using ZnO beds, prior to the syngas entering the FT reactors.

Sulfur Recovery - Receives sour (H_2S-rich) gas streams and converts H_2S to elemental sulfur and any NH_3 to N_2 in a three-stage Claus unit. Tail gas is treated in a SCOT unit prior to discharge through a catalytic incinerator to the stack.

Sour Water Stripping - Strips the water used for syngas wet scrubbing. Wastewater is sent to waste water treatment and the stripped gas to the sulfur plant.

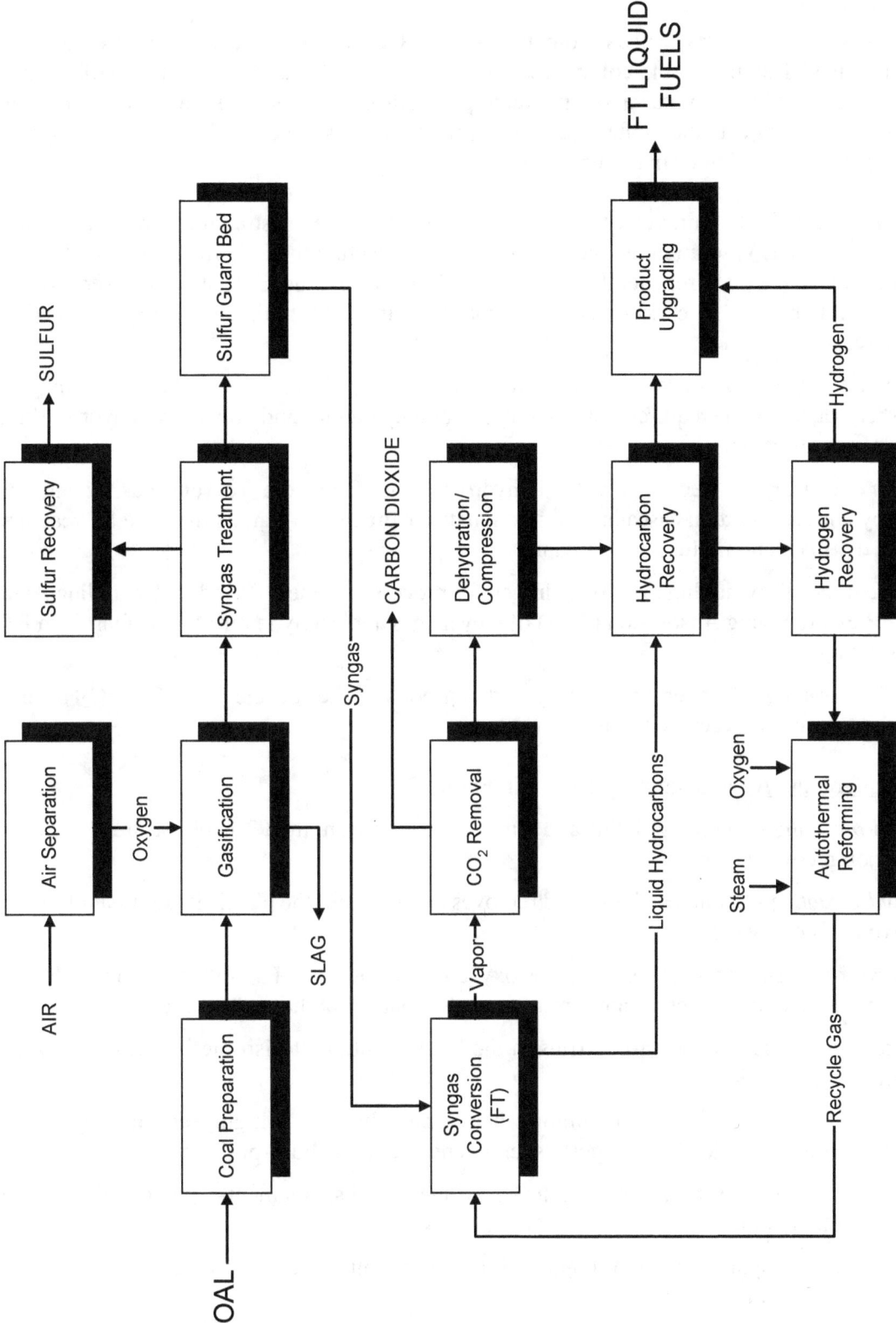

Figure 2. **Block Flow Diagram of Coal Liquefaction Process**

19

FT Conversion Area

Syngas Conversion - Converts syngas from the Syngas Generation Area and recycle gas into hydrocarbons using FT slurry bubble-column reactors; includes facilities for pretreatment of the iron FT catalyst, removal of the separate vapor and liquid phases from the reactor, separation and recycle of the catalyst withdrawn with the molten wax phase (physical and supercritical extraction), disposal of spent catalyst, and addition of make-up catalyst.

CO_2 Removal - Selectively removes CO_2 from the FT overhead vapor stream (recycle gas) using proprietary amine (MDEA) solution; includes absorber for contacting the CO_2-rich syngas with CO_2-lean solvent, and stripper for regenerating solvent. A portion of the CO_2 stream is sent to the gasification plant to be used as carrier gas for the coal feed and the remainder is directly vented to the atmosphere.

Dehydration & Compression - Pressurizes and removes moisture from the recycle gas leaving the amine absorber, satisfying the requirements for recycle loop hydraulics and downstream hydrocarbon recovery at low temperatures.

Hydrocarbon Recovery - Recovers C3-C4 hydrocarbons from the recycle gas, using an ethylene/propylene refrigeration cascade, and fractionates hydrocarbon liquids from the FT reactors into naphtha, distillate and molten wax streams.

Hydrogen Recovery - Provides high-purity hydrogen for processes in the FT Product Upgrading Area by means of Pressure Swing Absorption (PSA) of recycle gas and catalytic reformer offgas from FT naphtha upgrading.

Autothermal Reforming - Converts remaining hydrocarbons in the recycle gas (CH_4, C_2H_4, and C_2H_6) back into syngas for recycle to the FT reactors.

FT Product Upgrading Area (details not shown in Figure 2)

Naphtha Hydrotreating - Saturates olefins and removes oxygen from the FT naphtha stream leaving the hydrocarbon recovery plant.

Distillate Hydrotreating - Saturates olefins and removes oxygen from the FT distillate stream leaving the hydrocarbon recovery plant.

Wax Hydrocracking - Saturates olefins, removes oxygen, and cracks the FT wax stream from the FT reactors and hydrocarbon recovery plant, producing additional naphtha and distillate.

C5/C6 Isomerization - Isomerizes n-paraffins in the light naphtha into iso-olefins with improved gasoline-blending properties.

Catalytic Reforming - Converts the remaining heavy naphtha into a highly aromatic gasoline component with improved blending properties, and generates a medium-purity hydrogen offgas.

C4 Isomerization - Isomerizes n-butane from the FT synthesis and supplemental, purchased n-butane to isobutane for alkylation.

C3/C4/C5 Alkylation - Synthesizes additional high-quality gasoline blendstock from isobutane and C3/C4/C5 olefins from the FT process.

Saturate Gas Plant - Processes and separates offgas from various sources within the FT Product Upgrading Area producing LPG for sale, butanes for isomerization/alkylation and additional plant fuel gas.

<u>Offsites</u> (not shown in Figure 2)

Relief & Blowdown - Collection and flaring of relief and blowdown discharges from all applicable plants; includes two flare systems, one for hydrocarbon containing discharges and a secondary flare for discharges containing H_2S.

Tankage - Storage and delivery of products, intermediates and chemicals.

Interconnected Piping System - Includes process and utility piping between process plants and offsites.

Product Shipping - Provides the pipeline and metering system for the delivery of final FT naphtha and distillate products to customers.

Tank Car/Truck Loading - Provides pumping and loading/off-loading facilities for by-products (propane and sulfur) shipped and catalysts and chemicals received by tank car or tank truck.

Coal Ash Disposal - Transports coal ash and slag via conveyor back to coal mine for disposal as land reclamation.

Catalyst & Chemicals Handling - Provides storage and handling for catalysts and chemicals used in all plants.

Electrical Distribution System - Receives power from across-the-fence utility substations and distributes electricity to all applicable plants.

Steam & Power Generation - Manages and distributes all steam used and generated in all applicable plants and provides for excess steam for on-site power generation.

Raw, Cooling & Potable Water - Provides water treatment for make-up water withdrawn form nearby lakes or rivers, and distributes cooling and potable water to all applicable plants; includes cooling tower.

Fire Protection System - Provides fire protection and control systems for all facilities, structures and equipment.

Sewage & Effluent Water Treatment - Treats all wastewaters, including coal storage pile runoff, oily wastewater, process wastewater, solids de-watering and sanitary sewage.

Instrument & Plant Air Facilities - Provides instrument and utility air to all applicable plants and support facilities.

Purge & Flush Oil System - Delivers light and heavy flush oil for pump seal flushing and instrument purging.

Solid Waste Management - Disposes of wastes from raw, cooling and potable wastewater treatment.

General Site Preparation - Leveling and grading greenfield construction site; includes improvements such as roads, fencing, drainage, and placement of load-bearing fills, pilings and building foundations.

Buildings - Construction of all facilities onsite.

Telecommunications Systems - Provides telecommunications services for construction and operation of facility.

Distributed Control Systems - Provides control systems for monitoring and operating all applicable plant operations.

Options 2-4 involve variations of this basic design. For Option 2 - Illinois No. 6 Coal with Alternate ZSM-5 Product Upgrading, the following modifications are included:

Syngas Conversion - ZSM-5 reactors are provided directly downstream of the FT reactors to convert all overhead product leaving the FT reactors into a premium gasoline blending component. In turn, this simplifies the design of the FT Product Upgrading Area. *Naphtha Hydrotreating, Distillate Hydrotreating C5/C6 Isomerization, and Catalytic Reforming* processes are not required.

The only modifications to the basic design required for Option 3 - Illinois No. 6 Coal with Conventional Product Upgrading (maximum gasoline & chemicals production) are in the FT Product Upgrading Area. *Wax Hydrocracking* is not included, and the following processes have been added:

Fluid Catalytic Cracking - Cracks the FT wax stream from the FT reactors and hydrocarbon recovery plant, producing additional naphtha, light olefins for alkylation and ether synthesis, and a small quantity of distillate.

Ether Synthesis - Synthesizes gasoline blending ethers from C4 and higher iso-olefins using MTBE and TAME process units.

Only plant-specific modifications and changes to operating conditions (primarily in the Syngas Generation Area) are required for Option 4 - Wyoming Powder River Basin Coal with Conventional Product Upgrading:

Acid Gas Removal - Because of the high CO_2/H_2S ratio in the syngas, the amine absorption system is replaced with a Rectisol (methanol) wash system.

Raw, Cooling & Potable Water - This plant was redesigned by Bechtel for zero discharge to conserve water usage in an arid climate (Wyoming).

3.2.2 Biomass Based Design

The design considered in Option 5 - Biomass with Conventional Product Upgrading and Once-Through Power Generation is shown in the block flow diagram in Figure 3. This design is for a much smaller plant having only a single gasification train and only producing 1,156 bpd of FT liquid products versus the roughly 50,000 bpd produced in the previous designs. A breakdown of the various process plants appearing in the biomass design that differ from Option 1 is given below:

Syngas Generation Area

Wood Receiving & Storage (not shown in Figure 3) - Replaces coal receiving and storage.

Wood Preparation - Replaces coal preparation; dries wood chips prior to gasification.

Indirect Gasification - Feeds dried wood chips to a low-pressure, indirectly heated gasifier for gasification; includes char combustor and sand recirculation loop.

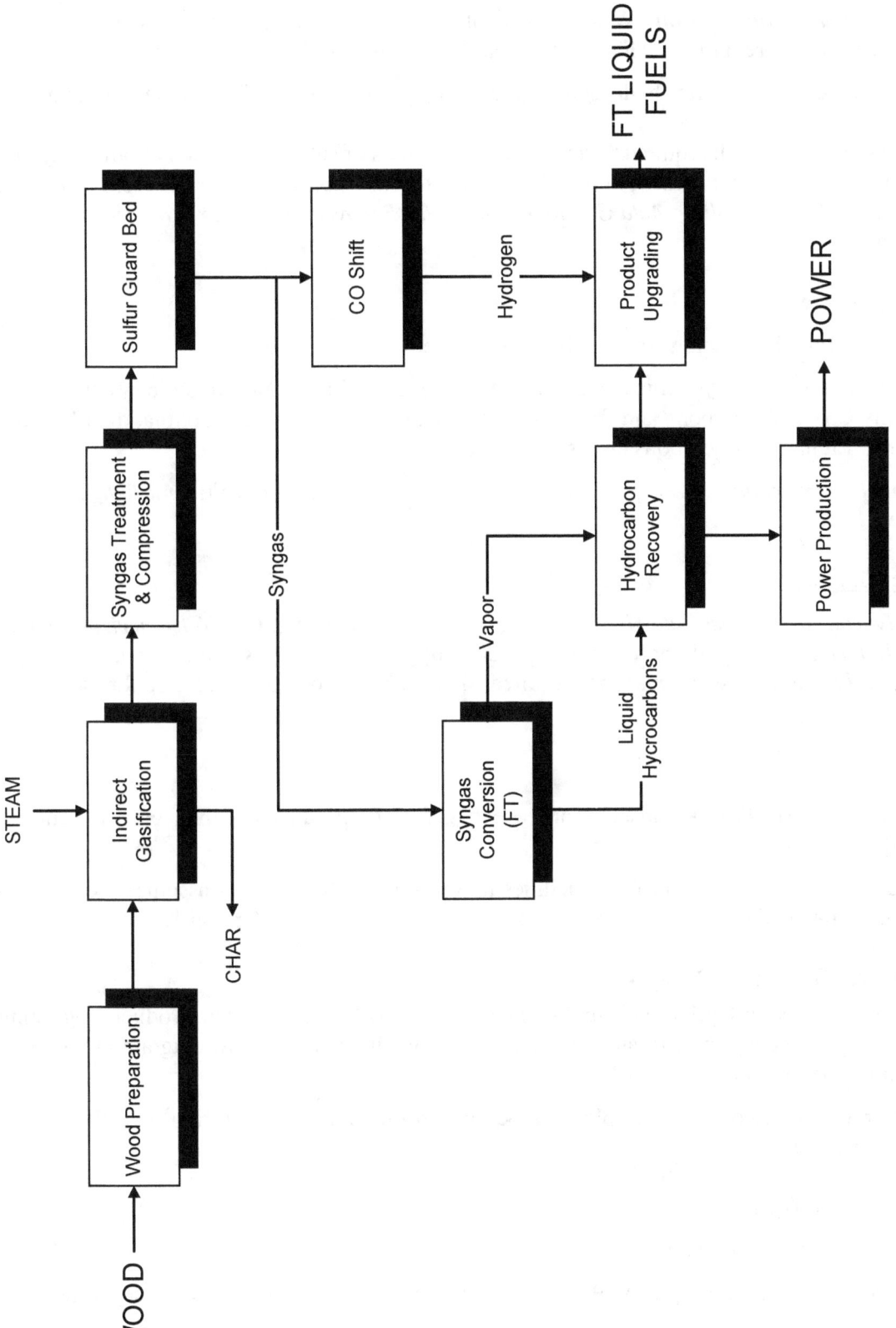

Figure 3.   Block Flow Diagram of Biomass Liquefaction Process

23

Syngas Treatment & Compression - Washes and cools syngas in a spray column before compressing syngas up to pressures required for FT synthesis and power generation.

CO Shift - Produces and purifies hydrogen from treated syngas used for FT product upgrading.

The *Sulfur Guard Bed* is still required to remove trace amounts of sulfur compounds from the syngas (small amounts of sulfur are present in the biomass feed). *Air Separation, Syngas Wet Scrubbing, COS Hydrolysis & Gas Cooling, Acid Gas Removal, Sulfur Recovery, and Sour Water Stripping* are not required.

FT Conversion Area

Syngas Conversion - FT reactors and catalyst systems remain unchanged.

Hydrocarbon Recovery - Cryogenic design has been replaced with a non-cryogenic system, which recovers only C5+ hydrocarbons and fractionates hydrocarbon liquids into naphtha, distillate and wax streams. Lighter hydrocarbons are used as fuel gas.

CO_2 Removal, Dehydration & Compression, Hydrogen Recovery, and Autothermal Reforming are not required.

FT Product Upgrading

Naphtha Hydrotreating, Distillate Hydrotreating, Wax Hydrocracking, C5/C6 Isomerization, and Catalytic Reforming are still included for product upgrading. *C4 Isomerization, C3/C4/C5 Alkylation, and Saturate Gas Plant* are not required, since light hydrocarbons are used for fuel in this design.

Offsites

Combined-Cycle Power Plant - Consumes all the excess fuel gas produced by the facility to generate electric power for sale.

Bechtel did not redesign any other offsite facilities for this option. Rather, they assumed these would remain approximately the same and prorated requirements using design Option 1.

3.2.3   Natural Gas Based Designs

The design considered in Option 6 – Pipeline Natural Gas with Conventional Product Upgrading (1990 technology - maximum distillate production) is shown in the block flow diagram in Figure 4. This design is very similar to Option 1.

A breakdown of the various process plants appearing in this natural gas design that differ from Option 1 is given below:

Syngas Generation Area

 Natural gas is supplied by pipeline.

Air Separation - Provides high-purity (99.5%) oxygen for POX using cryogenic air separation.

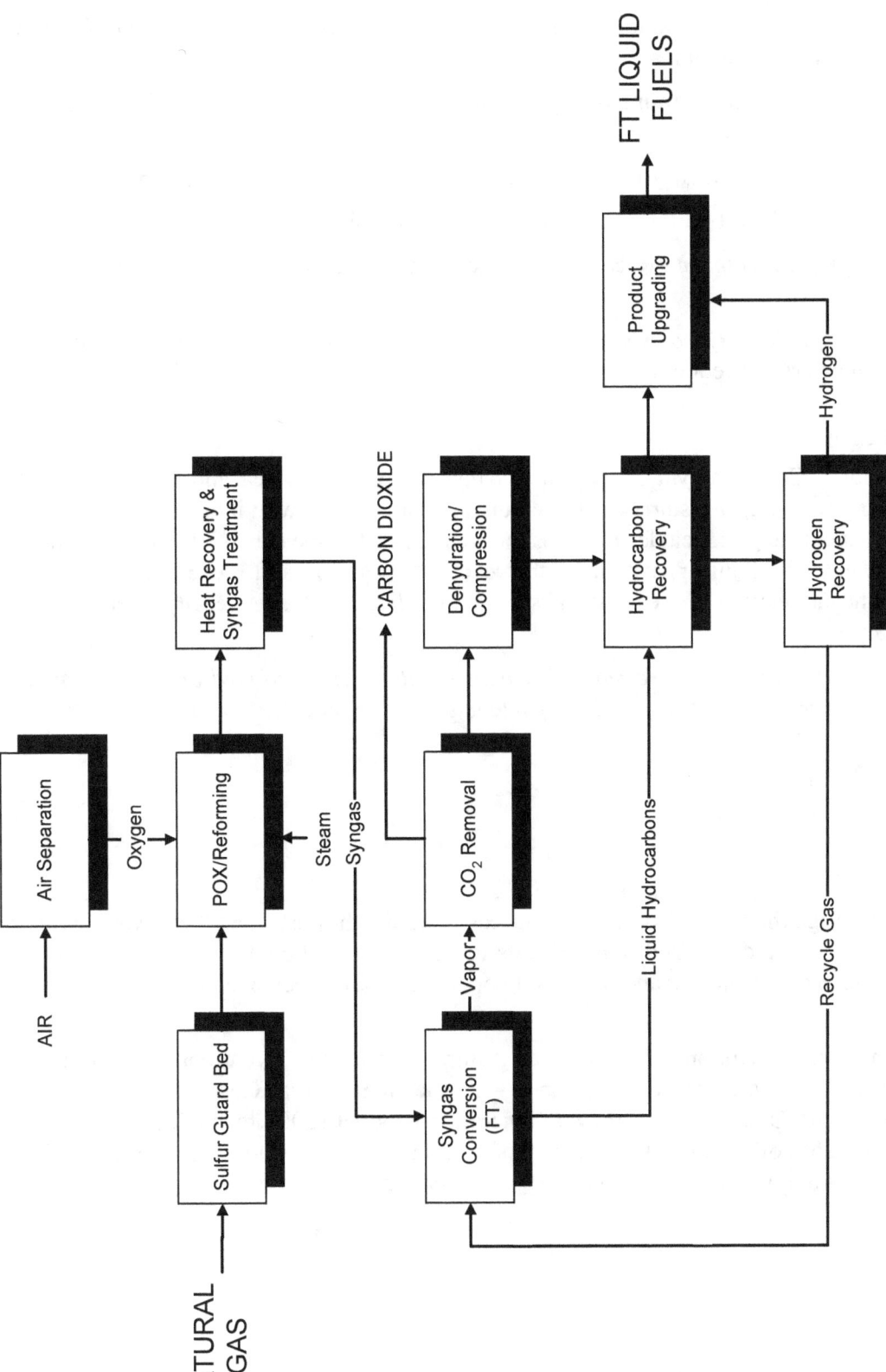

Figure 4. **Block Flow Diagram of Gas-To-Liquid Process - Old Design**

25

Sulfur Guard Bed - Removes trace amounts of sulfur compounds from the natural gas prior to the POX and steam reforming reactors.

POX/Reforming includes parallel trains of these units to achieve desired H_2 to CO ratio for FT synthesis:

POX - Partially oxidizes natural gas to syngas using oxygen form the air separation plant.

Steam Reforming - Catalytically reforms natural gas to syngas using steam.

Heat Recovery & Syngas Treatment - Recovers heat and scrubs traces of particulates from the cooled syngas.

Syngas Wet Scrubbing, COS Hydrolysis & Gas Cooling, Acid Gas Removal, Sulfur Recovery, and Sour Water Stripping are not required.

FT Conversion Area

Syngas Conversion - Converts syngas from the Syngas Generation Area and recycle gas into hydrocarbons using two-stage FT slurry bubble-column reactor system with interstage hydrocarbon removal from the overhead gas; includes facilities for pretreatment of the cobalt FT catalyst, removal of the separate vapor and liquid phases from the reactor, separation and recycle of the catalyst withdrawn with the molten wax phase (physical separation), disposal of spent catalyst, and addition of make-up catalyst.

CO_2 Removal, Dehydration & Compression, Hydrocarbon Recovery, and Hydrogen Recovery are still required. *Autothermal Reforming* of the recycle gas is not included.

FT Product Upgrading

Upgrading is identical to Option 1.

Offsites

Bechtel did not redesign the offsite facilities for this case. Again, they assumed these would remain approximately the same and prorated requirements using design Option 1. All offsites that are required solely due to coal handling and processing operations have been excluded.

The designs considered in Option 7 – Associated Natural Gas with Conventional Product (2000 technology - minimum upgrading) and Option 8 – Associated Natural Gas with Conventional Product Upgrading and Once-Through Power Generation Product (2000 technology - minimum upgrading) are variations of the block flow diagram shown in Figure 5. A breakdown of the various process plants appearing in these natural gas designs is given below:

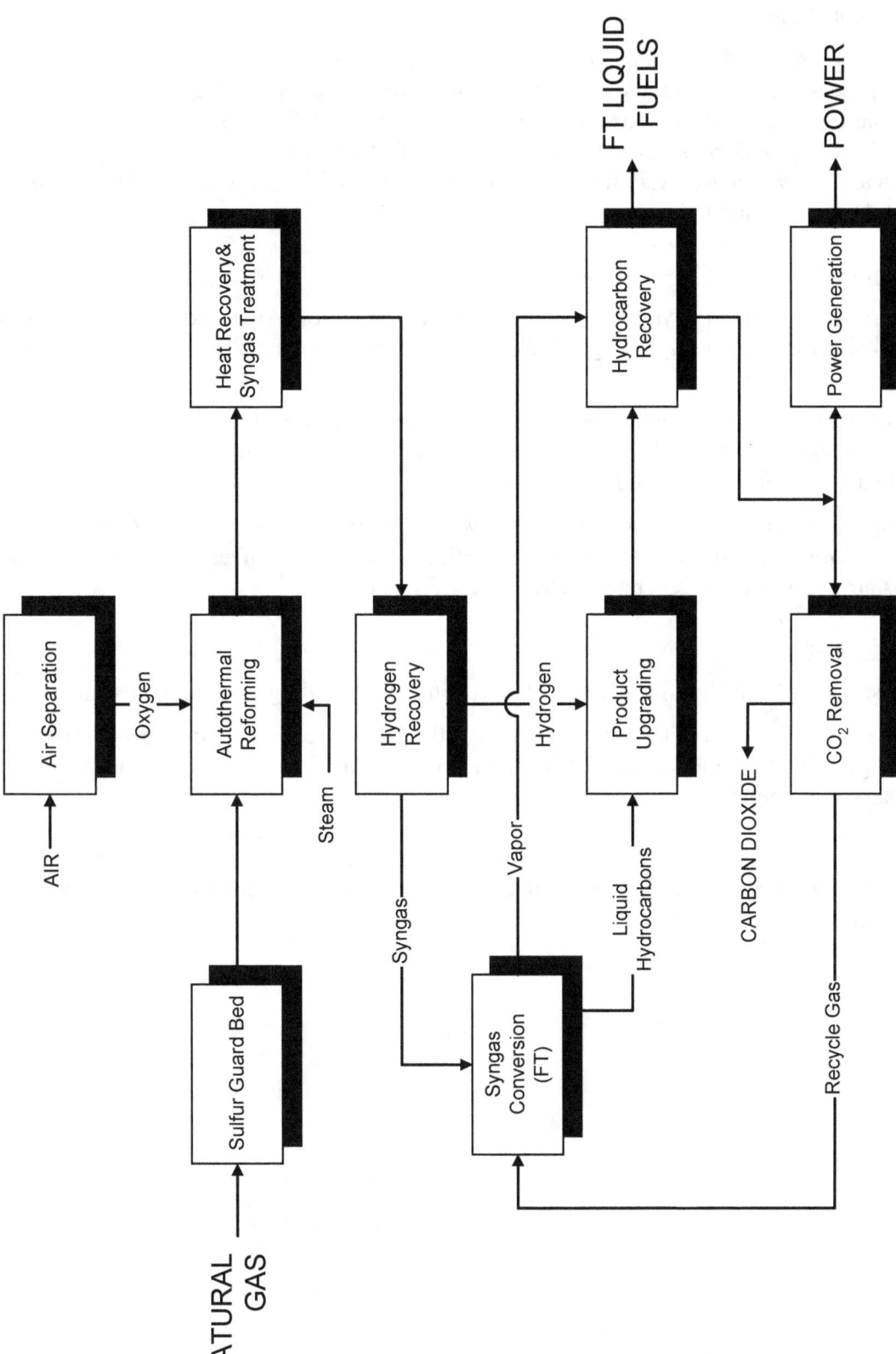

Figure 5. Block Flow Diagram of Gas-To-Liquid Process - New Design

27

Syngas Generation Area

Natural gas feed is *associated gas* from oil field production, which has been processed in an upstream gas processing plant to remove sour gas (H_2S), some natural gas liquids (C4s) and all natural gasoline (C5+ liquids). It contains significant amounts of CO_2. *Autothermal Reforming* replaces combined *POX/Reforming* to achieve desired H_2 to CO ratio for FT synthesis; requires both oxygen and steam. *Hydrogen Recovery* has been moved upstream of the FT reactors. All remaining processes are the same as in Option 6.

FT Conversion Area

Syngas Conversion - Converts syngas from the Syngas Generation Area and recycle gas into hydrocarbons using redesigned single-stage FT slurry bubble-column reactor system with cobalt FT catalyst.

Hydrocarbon Recovery - Coal design has been replaced with a non-cryogenic system, which recovers only C5+ hydrocarbons and fractionates hydrocarbon liquids into naphtha, distillate and wax streams. Lighter hydrocarbons are used as fuel gas.

CO₂ Removal has been moved to the syngas recycle loop in Option 7. *CO₂ Removal* and *Dehydration & Compression* are not required in Option 8, where unconverted syngas and C4-hydrocarbons are being used to generate electric power for sale.

FT Product Upgrading

Product upgrading has been significantly simplified (minimal upgrading case) and only includes:

Wax Hydrocracking - Cracks the FT wax stream from the FT reactors and hydrocarbon recovery plant producing additional naphtha and distillate, transportable by conventional oil transportation systems, tankers and pipelines.

Offsites
Combined-Cycle Power Plant - Consumes all the excess syngas/fuel gas produced by the facility to generate electric power for sale.

3.2.4 Resource Consumption & Yields

The various designs described in the preceding sections differ in their degree of detail. While the early designs completed by Bechtel were based on detailed sizing and costing [10-13], later designs were based on Aspen process simulation models developed primarily to fit the original designs (with modifications for the different technology options under consideration) [8,14,15]. For all the designs, however, material and energy balances were reported, which allow emissions to be calculated. In no case were these FT plant designs rigorously optimized, either for return on economic investment or to minimize emissions. They represent, as a group, the best-expected practices for these technologies at the time of their design.

Material and energy balance data from the eight designs being considered in this LCI were used to generate the resource consumption and yield data presented in Table 3. The basis for these values is 1 bbl of FT C3+ liquid product (combined C3/C4 LPG plus gasoline/naphtha plus distillate) unless noted. Yields are presented on a volume basis (bbl/bbl FT C3+ products), a mass basis (ton/bbl), and an energy basis (MM Btu (LHV)/bbl). The thermal efficiencies (LHV basis) of the coal and biomass liquefaction designs range from 47-52%. The thermal efficiencies of the natural gas designs are somewhat higher at 57-59%. The carbon efficiencies of the coal and biomass designs range from 37-41%. The carbon efficiency for the pipeline natural gas design is 57% and for both associated natural gas designs is about 39%. The large difference between the natural gas designs is due to the 13% CO_2 in the associated gas.

In addition to the primary feedstocks (coal, biomass or natural gas), the conversion plants require ancillary feedstocks: butanes and methanol used in specific FT product upgrading steps, raw water make-up (e.g., river water), catalysts and chemicals, and in some cases purchased supplemental electric power. Catalysts and chemicals have been aggregated to show that the amounts of these materials used are small relative to the primary feedstocks (1-2 wt%). Emissions associated with the production and delivery of catalysts and chemicals to the FT plant have been ignored for the LCI.

In the designs without recycle (Options 5 and 8), considerable power is generated and sold. Emissions and resource consumption have been allocated to the power, based on thermal input to the power generation device (gas or steam turbine). Option 6 also generates a small amount of power, which is sold to the electric grid. The fractions of all resources, by-products or emissions allocated to the fuels products are listed in Table 3. These allocations are 32.6%, 97.4% and 79.0% for Options 5, 6 and 8, respectively. Option 5 primarily produces power from biomass gasification; a result of the high methane content of the syngas produced by the low-temperature BCL gasifier. This methane is not directly available for conversion to higher hydrocarbons by the FT synthesis, and would require the addition of a steam reforming step to produce additional syngas. Allocations to power produced, on a per kWh basis, are listed in Table 3 in square brackets.

Table 3: Resource Consumption and Yields for FT Production
(Per bbl of FT Liquid Product)

Design	Option 1	Option 2	Option 3	Option 4	Option 5[1]	Option 6[1]	Option 7	Option 8[1]
Feedstock	IL #6	IL #6	IL #6	Wyo. Coal	Biomass	Pipeline Gas	Assoc. Gas	Assoc. Gas
Upgrading	Maximum Distillate	Increased Gasoline	Maximum Gaso. & Chem.	Maximum Distillate	Fuels & Power	Maximum Distillate	Minimum Upgrading	Min. Upgrading & Power
Resources								
Coal or Biomass (MF ton)	0.3675	0.3661	0.3310	0.395	0.621 [0.00072]			
Natural Gas (Mscf)						8.927 [0.018]	10.305	10.325 [0.012]
Butanes (bbl)	0.062		0.093	0.062		0.008		
Methanol (bbl)			0.041					
Catalysts & Chemicals (lb)	13.52	15.44	na	15.71	na	0.13	na	na
Water Make-Up (gal)	286	285	279	196	541 [0.629]	455 [0.923]	114	91 [0.105]
Electric Power (kWh)[2]	25.79	24.87	24.87	42.12	-1781	-13.2		-230
Volume Yield (bbl)								
C3/C4 LPG	0.038	0.071	0.118	0.038		0.038		
Gasoline/Naphtha	0.474	0.616	0.708	0.474	0.330	0.379	0.313	0.312
Distillates	0.488	0.313	0.174	0.488	0.670	0.583	0.687	0.688
Mass Yield (ton)								
C3/C4 LPG	0.003	0.007	0.011	0.003		0.003		
Gasoline/Naphtha	0.060	0.077	0.089	0.060	0.042	0.048	0.038	0.038
Distillates	0.066	0.043	0.023	0.066	0.091	0.079	0.092	0.092
Slag (MF)	0.044	0.044	0.040	0.035	0.065			
Sulfur	0.011	0.011	0.010	0.002				
Energy Yield (MMBtu)								
C3/C4 LPG	0.135	0.262	0.422	0.134		0.134		
Gasoline/Naphtha	2.120	2.764	3.019	2.121	1.463	1.687	1.439	1.433
Distillates	2.500	1.611	0.862	2.498	3.427	2.979	3.495	3.494
Power[3]					10.128	0.128		1.309
Allocation to Fuels					32.6%	97.4%		79.0%
Thermal Efficiency (LHV)	50.4%	52.0%	47.4%	49.3%	51.0%	59.1%	57.3%	57.1%
Carbon Efficiency	40.1%	41.1%	37.7%	39.1%	37.2%	57.0%	39.3%	39.2%

1 Values in [ ] are allocations per kWh of electricity produced and sold. All other values are per bbl of FT liquid product.

2 Positive value is purchase, negative value is sale.

3 Energy content of fuel used to produce power for sale.

In addition to the primary FT liquid products, ancillary products are also produced. These include elemental sulfur and slag for the coal-based designs (Options 1-4). Sulfur is sold as a by-product; however, no emissions have been allocated to it. Slag is returned to the coal mine for land reclamation. The biomass design (Option 5) produces a char/sand mixture from the gasifier, which could conceivably be sold for road asphalt manufacture. Again, emissions have not been allocated to slag or char. Wastewater discharges are not a significant issue for an inventory of airborne emissions and have not been included in Table 3. They are significant outflows from the Illinois sited FT plants (Options 1-3, 5 and 6). The Wyoming sited F-T plant (Option 4) was designed for zero water discharge.

3.3   Emissions from FT Production

Air emissions are generated from several sources within a FT conversion plant: combustion, vents, and fugitive sources. The conceptual designs developed by Bechtel meet all applicable federal and state (Illinois & Wyoming) statutes at the time of the design for airborne emissions of SOx, NOx, CO, VOC, and PM, including U.S. EPA New Source Performance Standards (NSPS).

Combustion emissions are associated with the burning of fuels within the plant. The primary fuel used in the FT designs is fuel gas generated in the FT Conversion Area (purged recycle gas) and the FT Product Upgrading Area (offgas). This fuel gas is a medium-Btu gas (300-400 Btu/scf) containing H_2, CO, and C1-C4 hydrocarbons. Fuel gas is used in fired heaters to provide process heat, in boilers to raise steam and in gas turbines to generate electric power. CO_2 emissions from fuel gas combustion were calculated from a carbon balance around the FT plant. For the other combustion related emissions, factors compiled by the EPA for refinery fuel gas were employed (see Section 6). The accuracy of this calculation is uncertain, since refinery fuel gas is a high-Btu gas (1000+ Btu/scf) rich in C1-C4 hydrocarbons. Different burner designs for these fuels will affect relative emissions of criteria pollutants. Gas turbine emissions of CH_4, CO and VOCs are generally higher than those from fuel gas combustion in a fired heater or boiler, and NOx emissions are generally lower [20]. Since the bulk of the fuel gas is used in fired heaters and boilers, adjustments to these emissions have not been made. For Option 5, where biomass is gasified in an indirectly heated gasifier, biomass char is burned in a fluidized bed combustor. Significant emissions are expected from this source. When catalysts are periodically or continuously regenerated (e.g., fluid catalytic cracking in Option 3) similar emissions can occur. Insufficient information was available to estimate emissions from these sources. However, they may be significant sources, particularly of NOx, CO and PM emissions.

Incineration is also a source of combustion emissions. The FT plant designs include a flare system for combustion of offgas produced during the normal operation of the plant and during start-up, shutdown, and process upsets. Flare emissions of methane have been estimated based on data for U.S. refineries (5.5 g CH_4 per refined bbl) [21]. It was assumed that the FT plant is of the same degree of complexity as an average U.S. refinery but has been designed to minimize flaring and, therefore, emissions are only half those reported for the average U.S. refinery. This seems reasonable for Options 1-6, where FT product upgrading includes many major refinery processes. For the associated gas Options 7 and 8, minimal refinery upgrading has been included, and it has been further assumed that emissions might be half of those expected from the other designs. Options 1-4 include a sulfur recovery plant, which generates a tail gas stream containing trace amounts of volatile sulfur compounds (H_2S and COS). This stream is catalytically combusted and sent to a separate flare. SOx emissions have been estimated based on the reported composition of this stream.

Vent emissions are point source emissions from the direct venting of process and utility streams to the atmosphere. The most significant stream in this category, and the only one included in this inventory, is the high-purity CO_2 stream vented from the CO_2 removal plant. This is the major source of the GHG emissions from the FT conversion process.

Fugitive emissions are releases from leaking equipment (valves, pumps, etc.), storage tanks and waste water treatment facilities. Since the FT plant designs are for state-of-the art facilities, they have been designed to minimize fugitive emissions of criteria pollutants. Fugitive emissions of CH_4 have been estimated based on data for U.S. refineries. For state-of-the art FT conversion facilities, it has been assumed that these emissions are only half those reported for the average U.S. refinery (231 g CH_4 per refined bbl). Emissions of CO_2 are not currently regulated, and roughly 1% of the CO_2 generated in the FT process is emitted from fugitive sources, primarily wastewater treatment operations.

3.3.1 Emissions Inventory for FT Production

Table 4 contains the LCI for the conversion step in the FT fuel chain for the eight FT plant designs considered in this study. Emission sources included in the inventory are fuel gas combustion, incineration, flaring, direct and indirect venting of CO_2, and upstream emissions from all ancillary feedstocks to the processes. The emission factors used to estimate these emissions and sample calculations are given in Appendix A. Ancillary emissions are presented in Section 6.

The clear trend in Table 4 is that most emissions are higher for the coal and biomass designs relative to the gas-to-liquid designs. All of the coal-based designs purchase supplemental electric power, and emissions from upstream electricity generation account for much of the difference for criteria pollutants. Coal also contains significant levels of sulfur, which is removed at the liquefaction plant. Tail gas from this process accounts for some of the SOx emissions for these designs; however, the bulk of SOx emissions are from ancillary power generation. The natural gas and biomass feedstocks contain only trace amounts of sulfur, and no bulk removal of sulfur compounds from the syngas is required. However, wellhead gas can contain significant amounts of H_2S, which would be removed in a gas processing plant upstream of a GTL facility. The SOx emissions listed for Option 6 are ancillary emissions related to the production of butanes used in the FT upgrading step.

Options 5 and 8 require special comment. Both produce significant excess power for sale. In this study, emissions were allocated between power and fuels in order to make comparisons between different design options. Table 5 contains the emissions for Options 5, 6 and 8 allocated to power on a per kWh of electricity produced and sold. *The procedure used for this allocation has a significant effect on the reported emissions per bbl of fuel produced.* This uncertainty is compounded by a lack of detailed information on fuel gas generation and consumption for some of the FT plant designs. Therefore, caution should be exercised when comparing the emissions from biomass liquefaction to coal liquefaction or to emissions from the various natural gas designs. Further work is needed to validate any benefits of co-producing fuels and power.

32

Table 4: Emissions Inventory for FT Production
(Per bbl of FT Liquid Product)

Design	Option 1	Option 2	Option 3	Option 4	Option 5*	Option 6*	Option 7	Option 8*
Feedstock	IL #6	IL #6	IL #6	Wyo. Coal	Biomass	Pipeline Gas	Assoc. Gas	Assoc. Gas
Upgrading	Maximum Distillate	Increased Gasoline	Maximum Gaso. & Chem.	Maximum Distillate	Fuels & Power	Maximum Distillate	Minimum Upgrading	Min. Upgrading & Power
CO_2 (g)	534311	526684	507159	575203	706987	119687	210964	92978
CH_4 (g)	58.55	51.14	64.40	87.27	12.97	8.45	4.77	4.79
N_2O (g)	2.16	1.91	2.11	2.85	16.50	1.60	2.02	3.17
SO_x (g)	197.64	190.73	193.85	298.04	0	0.06	0	0
NO_x (g)	89.08	72.07	98.31	118.82	523.90	51.93	64.15	100.51
CO (g)	15.66	11.73	18.02	19.09	127.23	12.61	15.58	24.41
VOC (g)	61.40	46.19	76.21	91.05	22.45	3.77	2.75	4.31
PM (g)	50.40	48.10	49.53	81.60	11.23	1.14	1.37	2.15

*Values reported only include allocation to fuel products.

33

Table 5: Emissions Inventory for Power Exported from FT Plants
(Per kWh of Electric Power)

Design	Option 5*	Option 6*	Option 8*
Feedstock	Biomass	Pipeline Gas	Assoc. Gas
Upgrading	Fuels & Power	Maximum Distillate	Min. Upgrading & Power
CO_2 (g)	822	243	107
CH_4 (g)	0.015	0.017	0.006
N_2O (g)	0.019	0.003	0.004
SOx (g)	0.000	0.000	0.000
NOx (g)	0.609	0.105	0.116
CO (g)	0.148	0.026	0.028
VOC (g)	0.026	0.008	0.005
PM (g)	0.013	0.002	0.002

*Values reported only include allocation to exported power.

3.3.2 Greenhouse Gases Emissions from FT Production

Greenhouse gas emissions for the FT designs have been compiled separately in Table 6. Emissions of CH_4 and N_2O have been converted to CO_2 equivalents using the GWPs in Table 1 for a 100-year time horizon. The GHG emissions in Table 6 have been broken up into the categories of vented gas, combustion and incineration flue gas, fugitive emissions and flaring, and ancillary emissions. GHG emissions are clearly dominated by direct CO_2 emissions; CH_4 and N_2O emissions account for less than 1% of total GHG emissions from the FT plants.

For the coal-based designs, the largest single source of GHG emissions is CO_2 removal (vented gas), followed by combustion of flue gas. Incineration flue gas and ancillary emissions are of roughly the same magnitude for the Illinois No. 6 coal designs. Incineration flue gas emissions are much smaller for the Powder River Basin coal. This is due to the higher sulfur content of Illinois coal versus Wyoming coal, which results in a larger gas stream being incinerated. However, overall GHG emissions are higher for the Wyoming coal and the biomass designs. This results from the high oxygen contents of these feedstocks (44 wt% for biomass and 17% for Wyoming subbituminous coal vs. 8% for Illinois #6 bituminous coal).

Table 6: GHG Emissions from FT Production
(Per bbl of FT Liquid Product)

Design	Option 1	Option 2	Option 3	Option 4	Option 5*	Option 6*	Option 7	Option 8*
Feedstock	IL #6	IL #6	IL #6	Wyo. Coal	Biomass	Pipeline Gas	Assoc. Gas	Assoc. Gas
Upgrading	Maximum Distillate	Increased Gasoline	Maximum Gaso. & Chem.	Maximum Distillate	Fuels & Power	Maximum Distillate	Minimum Upgrading	Min. Upgrading & Power
CO_2 – vented gas (g)	443800	441652	400060	440972	0	64289	94294	0
CO_2 – combustion flue gas (g)	47685	44538	65931	92081	706987	54565	115726	92978
CO_2 – incineration flue gas (g)	17803	17739	16037	5493	0	0	0	0
CO_2 – fugitive emissions (g)	5105	5081	4601	5126	0	643	943	0
CO_2 – ancillary sources (g)	19917	17675	20530	31531	0	191	0	0
CH_4 – combustion flue gas (g CO_2-eq)	15	12	14	15	225	22	28	43
CH_4 – fugitive & flaring (g CO_2-eq)	145	145	145	145	47	141	73	57
CH_4 – ancillary sources (g CO_2-eq)	1070	917	1193	1673	0	14	0	0
N_2O – combustion flue gas (g CO_2-eq)	331	266	328	334	5115	497	626	981
N_2O – ancillary (g CO_2-eq)	337	325	327	551	0	0	0	0
Total (g CO_2-eq)	536209	528350	509166	577921	712374	120361	211690	94060

* Values reported only include allocation to fuel products.

35

Natural gas, which is rich in hydrogen, does not produce as large a quantity of CO_2 during FT conversion (as can be seen by comparing the carbon efficiencies given in Table 3 for Option 6); and thus, has much lower GHG emissions than those from coal and biomass. Figure 6 clearly shows this effect for Options 1 and 6, which use different feedstocks (coal and natural gas) but produce the same FT products. Vented emissions of CO_2 are a smaller fraction of total GHG emissions for this natural gas design. This observation correlates well with the efficiencies of the two processes, 50% and 59% for Options 1 and 6, respectively. The large difference in GHG emissions between Options 6 and 7 is attributed to the high CO_2 content of the associated gas (13 vol%) versus the pipeline natural gas (less than 1%). There may be other small effects from the differences in the basic process designs. Option 8 would seem to indicate that GHG emissions could be greatly reduced by co-producing power. As was mentioned earlier, this may be an artifact of the allocation procedure used and requires further analysis. The fuels and power co-production designs do not contain a CO_2 removal step. Therefore, all CO_2 generated during the syngas generation and FT conversion steps is exhausted in the combustion flue gas streams.

No great differences exist between the emissions from the alternative upgrading Options 1, 2 and 3. Therefore, Option 1 will be used as the basis for Scenario 1 in the full GHG emissions inventory given in Section 7. Option 4, Wyoming coal, is the basis for Scenario 2; Option 5, biomass conversion, is the basis for Scenario 3; and Option 6, pipeline gas conversion, is the basis for Scenario 4. Option 7 is the basis for both Scenarios 5 and 6, which involve the conversion of stranded natural gas associated with oil production. Option 8 is used as the basis for the estimates made in the sensitivity analysis in Section 7.3 for the effects of co-production on GHG emissions.

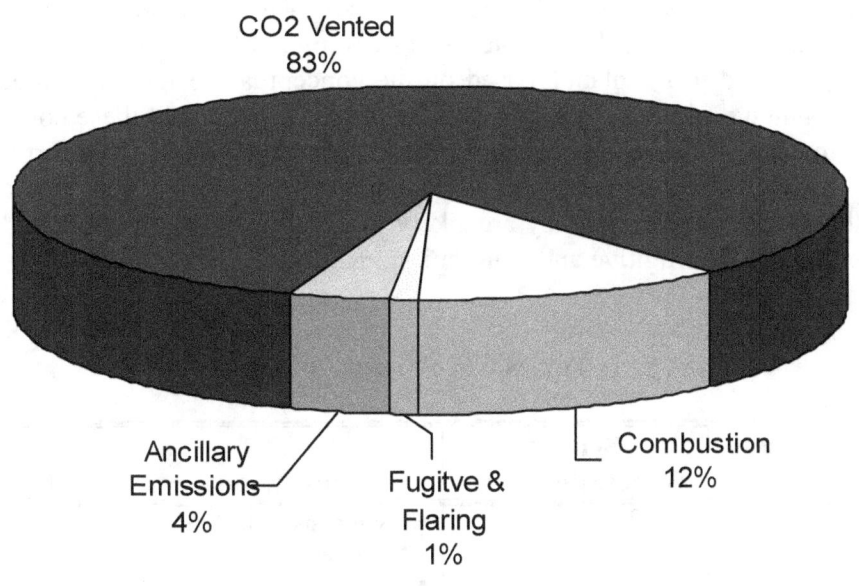

Design Option 1
536,209 g CO$_2$ -eq/bbl FT Product

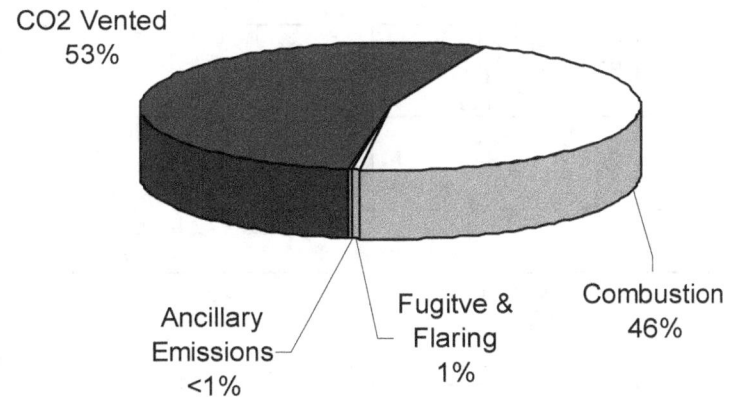

Design Option 6
120,362 g CO$_2$-eq/bbl FT Product

Figure 6.   Comparison of GHG Emissions Sources for FT Production

3.3.3 Air Toxics Checklist for FT Production

Some of the emissions that would arise from leaking equipment and process vents in FT plants are air toxics and hazardous air pollutants (HAPs). Releases of these compounds must be reported annually to the U.S. EPA. A checklist (Table 7) was compiled of compounds requiring reporting that are used or produced in FT plants, based on the conceptual designs described previously. Table 7 identifies which designs are affected and the possible sources of these compounds within the plant. While these compounds may be released as airborne emissions, no effort has been made to estimate what their emissions might be in an operating FT conversion facility. As stated previously, if these plants are built, they are likely to include state-of-the-art pollution control equipment, minimizing both fugitive and vent emissions.

Table 7: Air Toxics Checklist for FT Production

Chemical	Syngas Generation Area	FT Conversion Area	FT Product Upgrading Area
Aqueous Oxygenates: • Acetaldehyde • Formaldehyde • Methyl Ethyl Ketone		FT Synthesis - All Cases • Fe Catalyst • Trace from Cobalt Catalyst	
Aromatics: • Benzene • Toluene • Xylenes • Ethyl Benzene		• ZSM-5 Conversion - Option 2	• Cat Reforming - Options 1, 3-6 • Cat Cracking - Option 3
Sulfur Compounds: • Carbon Disulfide • Carbonyl sulfide	Coal - Options 1-4 • Gasification		
Acids: • Hydrochloric Acid	Coal - Options 1-4 Biomass - Option 5 • Gasification		
Olefins: • Ethylene • Propylene		FT Synthesis - All Cases • Fe Catalyst • Trace from Cobalt Catalyst	• Cat Cracking - Option 3
Alkane Solvents: • Hexane		FT Synthesis - All Cases	
Alcohols & Ethers: • Methanol • Methyl Tert Buytl Ether	• Rectisol Unit - Option 4		• Ether Synthesis- Option 3
Trace Elements: • Antimony, Arsenic, Barium, Beryllium, Boron, Cadmium, Chromium, Cobalt, Copper, Lead, Manganese, Mercury, Molybdenum, Nickel, Selenium, Vanadium	Coal - Options 1-4		

4. RESOURCE EXTRACTION

The three feedstocks considered in this analysis have quite different properties and are produced in very different ways: mining, farming and drilling. It is the relative proportions of carbon, hydrogen and oxygen in these resources and the size of the molecular structures present that give them their unique properties. Coal and biomass are solids composed of large molecules. Coals have molar hydrogen-to-carbon ratios less than 1 (0.8 for the coals considered here) and biomass has ratios between 1 and 2 (1.5 for the maplewood chips). However, during gasification, hydrogen reacts with oxygen in these feedstocks to produce H_2O. Thus, the effective hydrogen-to-carbon ratios of coal, and in particular biomass, can be much lower. Natural gas has a much higher hydrogen-to-carbon ratio of about 4. Most liquid hydrocarbons have a ratio of about 2. It is the relative deficiency or surplus of hydrogen in a feedstock, which most affects the severity of the operations necessary to convert the feedstock to liquid fuels. In turn, this affects the overall efficiency of FT conversion and the amount of CO_2 generated in the process.

4.1 Coal

Coals are classified according to their *rank*, which is defined based on the coal's fixed carbon, volatile matter, and heating value. In addition to these properties, the ash (mineral matter), moisture, sulfur, nitrogen and oxygen contents are also important. Sulfur and nitrogen contents are indicative of SOx and NOx emissions, which result from burning coal. The four major rankings used for coals are anthracite (high fixed carbon, low volatile matter, high heating value), bituminous, subbituminous and lignite (low fixed carbon, high volatile matter, low heating value). Rank is also indicative of the age of the coal seam from which the coal was mined, with lignite being the least advanced along the path to becoming anthracite coal. Bituminous coals, such as Illinois No. 6, are found in the eastern United States. Powder River Basin coal from Wyoming is typical of western subbituminous coals. The FT plant designs discussed in Section 3 were based on these two benchmark coals. These coals were selected for the conceptual designs because they are representative of the bulk of the coal used in the U.S. and because a considerable amount of information is available on them, including results from coal preparation and gasification tests. Analyses of Illinois No. 6 and Powder River Basin coal are given in Table 8.

Table 8: Ultimate Analyses of Coal and Biomass

	Illinois #6 Coal	Wyoming Coal	Maplewood Chips
HHV (M Btu/lb)	12.25	11.65	8.08
LHV (M Btu/lb)	11.95	11.20	7.72
	Wt. %	Wt. %	Wt.%
Moisture	9.41	44.9	61.0
Ash	11.49	8.71	0.50
C	71.01	67.84	49.54
H	4.80	4.71	6.11
N	1.40	0.94	0.10
S	3.19	0.58	0.02
Cl	0.10	0.01	0.00
O (by diff.)	8.01	17.21	43.73

4.1.1 Coal Mining and Post Mining Operations

Depending on local geological conditions, a number of options are available for coal mining. Economics dictate the method used to mine any given site, with the depth of the coal seam being a major factor. When a coal seam is near to, or breaks, the surface (i.e. *outcrops*), surface mining techniques are employed, such as *strip mining*. Western coals, such as Powder River Basin coal are primarily mined this way. Roughly 60% of the coal mined in the U.S. is surface-mined. When the coal seam lies sufficiently deep, underground mining techniques are employed. The two most common underground methods used in the U.S. are *room-and-pillar* and *longwall* mining. Longwall mining is the newer method and typically has economic, as well as other, advantages over traditional room-and-pillar mining. Eastern coals, such as Illinois No. 6, are often found in deeper seams, where both underground mining techniques are used. However, eastern coals are also surface-mined where possible. Other less common techniques are also still in use.

Underground mining involves excavating a number of shafts from the surface to the coal seam. These shafts may be vertical, horizontal or at some other angle depending on the topography of the mine site. Room-and-pillar and longwall mining differ by the methods and machinery used to remove the coal from the seam. In room-and-pillar mining, the coal is removed from two sets of corridors that advance through the mine at right angles to each other. The remaining, evenly spaced pillars of coal are left in place to support the overlying layers of rock. As much as half the coal in the seam is left in place for support. Even so, over long periods of time (decades to centuries), the mine will collapse, possibly causing surface subsidence. The machine used to remove coal in room-and-pillar mining is called a continuous miner. Mining using a continuous miner involves a series of operations: drilling, blasting, cutting, loading and hauling.

In longwall mining, three main corridors are first mined (using continuous miners) to form a large U-shaped passageway. The distance between the two parallel corridors is on the order of 100 to 200 meters. The "longwall" in the corridor perpendicular to these two corridors is mined continuously, using a longwall-mining machine. This machine, which has a movable roof support, advances as

40

it removes coal from the *coalface*. Behind it, the unsupported mine roof quickly collapses, resulting in controlled surface subsidence. The coal is transported by means of conveyors to either end of the longwall where it is hauled out of the mine. With longwall mining, no coal is left in the mined portion of the seam. Many of the other operations required in underground mining are similar for both room-and-pillar and longwall mining. They include providing rock dusting, water supply, ventilation, drainage, power supply, communications and lighting.

Because longwall mining is the most efficient and lowest cost option for underground mining and is gradually replacing the older room-and-pillar method, *only longwall mining has been considered as part of this emissions inventory*. Machinery for longwall mining operations includes the longwall unit, auxiliary continuous miners, shuttle cars, roof bolters, triple rock and trickle dusters, supply cars, conveyors, tracks, front-end loaders, bulldozers and other miscellaneous equipment and vehicles. Table 9 lists the resources consumed in longwall mining. Almost all equipment operated in underground coal mining is powered by electricity in order to maintain safe air quality within the mine. Limestone is used for rock dusting to reduce the risk of coal dust explosions, and water is used to cool and lubricate coal-cutting equipment.

Surface mining involves removing the overlying soil and rock, known as *overburden*, to expose the coal seam, removing and loading the coal for transportation, and replacement of the original soil and rock (land reclamation). Blasting and/or mechanical means are used to fracture the coal seam and any overlying layers of rock. Machinery required for surface mining operations includes stripping shovels, drills, bulldozers, coal shovels, coal haulers (trucks), front-end loaders with shovels, wheel tractor scrapers, road graders, forklifts, cranes, and other miscellaneous vehicles. Table 9 lists the resources consumed in surface mining. Since much of the equipment used in surface mining is mobile, distillate fuel is a significant source of power. This fuel can be assumed to be equivalent to high-sulfur, No.2 Diesel. Ammonium nitrate is the explosive most widely used in blasting.

Post-mining operations include coal preparation and storage before final shipment by train, truck or barge. Coal preparation involves size reduction of the mined coal to facilitate the separation of rock and mineral matter, known as ash, from the raw coal. This density-based separation is referred to as jig washing or cleaning. Other more advanced coal cleaning operations, such as heavy media separation and agglomeration, have been developed, but are not commonly used in the U.S. In addition to the cleaned coal, jigging produces a refuse stream of rock, mineral matter and very fine coal particles, which can be returned to the mine for use in land reclamation. Jigging also involves the use of large quantities of water, which can be recycled, but must be treated if discharged. Table 9 lists the resources consumed and refuse generated in a typical coal preparation operation.

Table 9: Resource Consumption for Coal Production*
(Per ton of MF Coal Produced)

	Illinois #6 Underground Mine	Illinois #6 Surface Mine	Wyoming Surface Mine
Electricity (kWh)	15.4	17.2	17.4
Distillate Fuel (gal)		0.840	0.085
Water Make-Up (gal)	62.6	46.1	44.7
Limestone (lb)	42.6		
Ammonium Nitrate (lb)		5.4	5.5
Refuse (ton)	-0.310	-0.310	-0.320

*Positive value is consumed, negative is produced; values based on [16,17].

Emissions associated with the production and delivery of limestone, ammonium nitrate, etc. to the coal mine have been ignored for the LCI. The amounts of these materials used are small relative to the coal produced (0.3-2.3 wt%).

4.1.2 Coalbed Methane

Methane (CH_4) is often found in association with coal seams, either absorbed in the seam or in pockets in adjacent rock strata. Methane, if it is not removed, is a significant mining safety hazard. The amount of methane that can be absorbed in coal is a function of coal rank. Higher rank coals tend to hold more methane than lower rank coals. This methane is released when the pressure within the coalbed is reduced, either through mining activity, or through natural erosion or faulting. Due to the latter, surface mined coals frequently do not have large quantities of methane associated with them.

Methane, if found in association with coal, may be released prior to mining using de-gasification wells. This methane can be used at the mine site to satisfy electricity needs or sold as pipeline-quality natural gas. It is frequently not recovered; however, and is vented or flared. This situation is beginning to change in the U.S. with more coalbed methane being recovered and utilized. In underground mines, ventilation systems are utilized to circulate air through the mine and maintain methane levels below explosion limits. Longwall mining can release large quantities of methane, since the associated subsidence releases gas from overlying rock strata. Methane remaining in the coal after it is brought to the surface is released during post-mining operations.

The methane emission factors used in this study for underground and surface mining of eastern and surface mining of western coal are listed in Table 10.

Table 10: Coalbed Methane Emissions*
(Per ton of MF Coal Produced)

	Illinois #6	Illinois #6	Wyoming

	Underground Mine	Surface Mine	Surface Mine
CH₄ (scf)	145	90	7.4
CH₄ (g)	2779	1725	142

*Based on [18].

4.1.3 Emissions Inventory for Coal Production

Table 11 contains the LCI for the coal production step in the FT fuel chain for the options: Illinois No. 6 coal - underground longwall mine, Illinois No. 6 - surface strip mine, and Powder River Basin coal - surface strip mine. Emissions sources included in the inventory are coalbed methane releases, ancillary electricity production, and ancillary diesel fuel production and use. The emissions factors used to estimate these emissions and sample calculations are given in Appendix A. Ancillary emissions are presented in Section 6. Table 12 contains the corresponding greenhouse gas emissions in CO_2 equivalency units.

Table 11: Emissions Inventory for Coal Production
(Per ton of MF Coal Produced)

		Illinois #6 Underground Mine	Illinois #6 Surface Mine	Wyoming Surface Mine
CO₂	(g)	10904	12272	12358
CH₄	(g)	2806	1754	172
N₂O	(g)	0.65	0.73	0.73
SOx	(g)	106.2	119.4	120.2
NOx	(g)	27.6	31.3	31.6
CO	(g)	3.2	3.67	3.7
VOC	(g)	27.8	31.2	31.4
PM	(g)	29.3	32.9	33.2

Table 12: Greenhouse Gas Emissions from Coal Production
(Per ton of MF Coal Produced)

	Illinois #6 Underground Mine	Illinois #6 Surface Mine	Wyoming Surface Mine
CO_2 (g)	10904	12272	12358
CH_4 (g CO_2-eq)	58928	36850	3618
N_2O (g CO_2-eq)	200	225	227
Total (g CO_2-eq)	70032	49348	16203

From Table 12, it is clear that coalbed methane emissions are a significant contributor to GHG emissions from coal mining. They are the dominant GHG emission for the Illinois underground and surface mining options. Only for the Wyoming surface mining option are coalbed methane emissions significantly smaller than emissions from mining operations.

The Illinois No. 6 underground mining and the Wyoming surface mining options are used as the basis for Scenarios 1 and 2, that are presented in Section 7.

4.2   Biomass

Biomass is a broad term used to refer to any material that is or was derived from plants and animals that were recently alive; this includes agricultural and animal products, forest and yard litter, wood waste from pulp mills, portions of landfill material, municipal solid waste, etc. These materials are renewable. They can be replaced by regrowth. However, this *regrowth must be accomplished in a sustainable way for the use of biomass to have a long-term benefit.*

The composition of biomass is highly variable. An examination of all possible sources for this feedstock is beyond the scope of this study. The only biomass feedstock considered in this study is maplewood, produced on a plantation as an energy crop specifically for use in the production of fuels and power. An analysis of this feedstock is given in Table 8.

4.2.1   Biomass Plantation Operations

The plantation is assumed to surround the biomass liquefaction plant, which has been sited in southern Illinois to be consistent with the eastern coal option. Best agricultural practices are assumed and there is a planned rotation of field plantings throughout the lifetime of the plantation. Fertilizer and herbicide use has been minimized. The average distance for the short-haul from the field to the plant is 17.25 mi. (27.6 km).

Energy is consumed and emissions released for each operation required to plant, grow and harvest the biomass. The equipment required per growing cycle includes plows, sprayers, spreaders, cultivators, tree fellers, bunchers, and chippers. Trucks are used to transport the chipped wood to the liquefaction plant. The major source of energy to operate this equipment is diesel fuel.

4.2.2 Emissions Inventory for Biomass Production

Table 13 contains the LCI for the biomass production step in the FT fuel chain. It is based on the LCA conducted by NREL for biomass-gasification combined-cycle power generation [19]. The biomass feedstock used in the NREL study was hybrid poplar. It has been assumed here that the emissions factors for maplewood cultivation and harvesting are the same as for hybrid-poplar wood. Because trees absorb CO_2 when they grow, the production of biomass results in a net removal of CO_2 from the atmosphere (the negative emission of CO_2 in Table 13). The effects of agriculture on soil and its ability to hold or absorb carbon are controversial, and it was assumed in the NREL study that agricultural best practices would not result in any net loss or gain of carbon in the soil. There is also great uncertainty as to emissions of CH_4 and N_2O during agriculture. NREL's study assumes only modest emissions of these gases from the soil.

Emission sources for biomass production were discussed in the previous section. The values given in Table 13 are aggregated for all sources associated with cultivation and harvesting, including ancillary feedstocks and short-haul transportation of the biomass from the fields to the FT conversion facility by diesel truck. Table 14 contains the corresponding greenhouse gas emissions in CO_2 equivalency units.

Table 13: Emissions Inventory for Biomass Production*
(Per ton of MF Biomass Produced)

		Feedstock Sequestering	Cultivation & Harvesting	Local Transportation	Total
CO₂	(g)	-1648273	52333	10162	-1585778
CH₄	(g)		8.3	0.39	8.7
N₂O	(g)		16.9	0.40	17.3
SOx	(g)		na	na	Na
NOx	(g)		307	49.4	356.4
CO	(g)		124	19.9	144
VOC	(g)		129.3	14.7	144
PM	(g)		na	na	Na

*Based on [19].

Table 14: Greenhouse Gas Emissions from Biomass Production
(Per ton of MF Biomass Produced)

	Feedstock Sequestering	Cultivation & Harvesting	Local Transportation	Total
CO_2 (g CO_2)	-1648273	52333	10162	-1585778
CH_4 (g CO_2-eq)		175	8.2	183
N_2O (g CO_2-eq)		5239	124	5363
Total (g CO_2-eq)	-1648273	57747	10294	-1580232

Plantation biomass is the basis for Scenario 3 of the full emissions inventory presented in Section 7.

4.3   Natural Gas

Natural gas occurs either separately from, or in association with, petroleum or coal. Methane (CH_4) is the major constituent, but other hydrocarbons such as ethane (C_2H_6), propane (C_3H_8), butanes (C_4H_{10}), and heavier (C5+) may also be present, especially when the gas is found in association with oil. The FT plant designs discussed in Section 3 considered two gas compositions. These are given in Table 15. The associated gas composition is typical of the gas produced along with Alaska North Slope oil. It contains 13% CO_2, negligible H_2S, and has been processed to remove and recover C5+ hydrocarbons. The composition of associated gas can vary considerably from location to location. The second composition given in Table 15 is for pipeline quality gas.

Table 15: Composition of Associated & Pipeline Natural Gas*

	Associated Gas	Pipeline Gas
HHV (Btu/scf)	925.3	1002.5
LHV (Btu/scf)	835.4	904.6
	Vol. %	Vol. %
Methane	76.2	94.7
Ethane	6.4	3.2
Propane	3.2	0.5
Isobutane	0.3	0.1
n-Butane	0.8	0.1
C5+	0.1	0.1
CO_2	12.6	0.7
H_2S	-	-
N_2	0.4	0.6

*Based on [9,13].

4.3.1 Oil & Gas Production Operations

Natural gas is produced from natural gas production wells or as associated gas from oil production wells. Natural gas is also produced from coalbed methane recovery wells, which have not been considered here. In either case, a field separation unit is used to separate produced gas, liquid hydrocarbons and liquid water. In a true gas field, the amount of liquid hydrocarbons produced is very small, and the liquid hydrocarbon mixture is referred to as *field condensate*. Gas from the field separators is gathered by a field pipeline network and fed to a gas processing plant. The purpose of the gas processing plant is to remove impurities in the gas, such as CO_2 and H_2S, and to recovery C3+ hydrocarbons. Removal of CO_2 and H_2S is referred to *gas sweetening*, and recovery of hydrocarbon liquids is referred to *gas conditioning*. Gas leaving the gas plant is of pipeline quality and is transported long distances to markets remote from the field in high-pressure natural gas transmission pipelines or liquefied cryogenically and shipped in LNG (liquefied natural gas) tankers. In oil fields, the gas may be re-injected into the reservoir to maintain pressure and enhance oil recovery. Ethane recovered from the gas may be sold as a petrochemical feedstock for producing ethylene or used as gas plant fuel. Propane, butanes and higher hydrocarbons recovered at the gas plant are referred to as natural gas liquids (NGLs). All are used as petrochemical feedstocks. Propane is also sold as LPG (liquefied petroleum gas) which is used as a fuel. Butanes are blended or converted into gasoline components, and C5+ liquids, referred to as *natural gasoline*, are also blended into gasoline.

4.3.2 Emissions Inventory for Natural Gas Production

Table 16 contains the LCI for the natural gas production step in the FT fuel chain. Emissions sources included in the inventory are natural gas venting and flaring, gas plant fuel combustion, and fugitive emissions. For pipeline natural gas, emissions for transportation and distribution are also included. It has been assumed that natural gas is the sole source of process fuel and power at the production site. Emissions of SOx for associated gas is negligible, since the composition of gas used (see Table 15) contains no sulfur compounds. This is not typical, as can be seen from the SOx value reported in Table 16 for the pipeline gas option. Table 17 contains the corresponding greenhouse gas emissions in CO_2 equivalency units.

Table 16: Emissions Inventory for Natural Gas Production*
(Per Mscf of Natural Gas Produced)

		Associated Gas	Pipeline Gas
CO_2	(g)	4427	6364
CH_4	(g)	22.8	69
N_2O	(g)	0.15	0.21
SOx	(g)	na	0.21
NOx	(g)	33.7	48.4
CO	(g)	8.2	11.8
VOC	(g)	53.6	77
PM	(g)	0	0

*Based on [20,21].

Table 17: Greenhouse Gas Emissions from Natural Gas Production
(Per Mscf of Natural Gas Produced)

	Associated Gas	Pipeline Gas
CO_2 (g CO_2)	4427	6364
CH_4 (g CO_2-eq)	478.8	1449
N_2O (g CO_2-eq)	45.3	65
Total (g CO_2-eq)	4951	7878

The difference in the emissions for pipeline versus associated gas is attributed to gas transportation and distribution. Pipeline gas is used as the basis for Scenario 4, and associated gas as the basis for Scenarios 5 and 6 in the full emissions inventory presented in Section 7.

5. TRANSPORTATION & DISTRIBUTION

The various scenarios considered for this inventory involve moving feedstocks and products over long distances. The means of transportation depends on the starting and ending point. All scenarios involve multiple transportation steps. To standardize comparisons, all the scenarios excluding Scenario 6, assume the end-use of the FT fuel occurs in the vicinity of Chicago, IL.

5.1   Transportation Modes & Distances

Scenarios 1 (Illinois No. 6 coal), 3 (biomass), and 4 (pipeline gas) all use southern Illinois as the location of the FT plant. The U.S. Midwest is a reasonable location for the future siting of coal liquefaction plants, as well as, biomass conversion plants. The high cost of pipeline gas makes Scenario 4 unlikely; however, it has been included to allow comparisons to be made between the different feedstocks on a consistent basis. The ultimate source of the pipeline natural gas has not been identified; however, a generic gas pipeline transmission step has been lumped into the emissions factor reported for pipeline natural gas production (see Tables 16 and 17, previous section).

The FT fuels produced in southern Illinois are shipped by pipeline to the Chicago area and distributed to local refueling station by tank truck. Scenario 2 assumes a Wyoming location for the FT plant, again with products shipped by pipeline to the Chicago area for distribution. Scenario 5 is based on the conversion of stranded, associated gas in Venezuela. Transportation of the FT fuels produced in Venezuela is by tanker to the U.S. Gulf Coast, followed by pipeline transmission to the Chicago area. While a small quantity of Alaska North Slope (ANS) crude finds its way to the Midwest every year, it is unlikely that substantial quantities of ANS crude or GTL would be refined and marketed there due to cost and logistic issues. Scenario 6 is based on FT production on the North Slope of Alaska (to monetize stranded gas reserves). The FT fuels produced are transported via the Trans-Alaska pipeline to Valdez, transferred to a tanker, and transported to the U.S. West Coast, where they are refined/blended into fuels for distribution in the San Francisco Bay area.

Energy usage for different modes of transportation is listed in Table 18. Mileage for the different transportation routes considered was estimated using standard atlases and is listed for the different scenarios in Tables 19-22.

Table 18: Energy Consumption for Different Modes of Transportation*
(Per ton-mile Transported)

Truck	Tanker	Tank Car	Pipeline
Btu	Btu	Btu	kWh
1900	408	516	0.0352

*Based on [20,21].

5.2 Emissions Inventory for Transportation & Distribution

Tables 19-22 contain the LCIs for the various transportation scenarios considered. Emissions sources included in the inventories are the combustion of the fuel used for each transportation step and upstream emissions associated with producing this fuel. Electricity is used to power pipeline pumps. Distillate fuel oil (DFO) is used for tank trucks, and residual fuel oil (RFO) for tankers. The emissions factors used to estimate these emissions and sample calculations are given in Appendix A. Ancillary emissions are presented in Section 6. Table 23 contains the corresponding greenhouse gas emissions in CO_2 equivalency units for all scenarios considered.

Table 19: Emissions Inventory for Transportation Scenarios 1, 3 & 4
(Per gal of FT Fuel Transported)

Transportation Mode		Truck	Tanker	Pipeline	Total
Southern Illinois to Chicago		DFO	RFO	Electricity	
Miles		60	0	200	260
CO_2	(g)	28.29	0	5.00	33.3
CH_4	(g)	0.0015	0	0.0124	0.0139
N_2O	(g)	0.0009	0	0.0003	0.0012
SOx	(g)	0.1389	0	0.0487	0.1876
NOx	(g)	0.1223	0	0.0185	0.1408
CO	(g)	0.1638	0	0.0059	0.1697
PM	(g)	0.0235	0	0.0134	0.0369
VOC	(g)	0.0011	0	0.00013	0.0012

Table 20: Emissions Inventory for Transportation Scenario 2
(Per gal of FT Fuel Transported)

Transportation Mode		Truck	Tanker	Pipeline	Total
Wyoming to Chicago		DFO	RFO	Electricity	
Miles		60	0	1000	1060
CO_2	(g)	28.29	0	25.00	53.30
CH_4	(g)	0.0015	0	0.0619	0.0634
N_2O	(g)	0.0009	0	0.0014	0.0023
SOx	(g)	0.1389	0	0.2434	0.3824
NOx	(g)	0.1223	0	0.0923	0.2147
CO	(g)	0.1638	0	0.0296	0.1934
PM	(g)	0.0235	0	0.0672	0.0907
VOC	(g)	0.0011	0	0.00067	0.0017

Table 21: Emissions Inventory for Transportation Scenario 5
(Per gal of FT Transported)

Transportation Mode		Truck	Tanker	Pipeline	Total
Venezuela to Chicago		DFO	RFO	Electricity	
Miles		60	2000	1200	3260
CO_2	(g)	28.29	218	30.00	276.23
CH_4	(g)	0.0015	0.2897	0.0742	0.3654
N_2O	(g)	0.0009	0.0050	0.0017	0.0076
SOx	(g)	0.1389	2.7352	0.2921	3.1663
NOx	(g)	0.1223	0.7158	0.1108	0.9489
CO	(g)	0.1638	0.1246	0.0355	0.3239
PM	(g)	0.0235	0.1652	0.0806	0.2693
VOC	(g)	0.0011	0.1077	0.00081	0.1096

Table 22: Emissions Inventory for Transportation Scenarios 6
(Per gal of FT Fuel Transported)

Transportation Mode		Truck	Tanker	Pipeline	Total
ANS to San Francisco		DFO	RFO	Electricity	
Miles		60	4130	800	4990
CO_2	(g)	28.29	450	20	498.32
CH_4	(g)	0.0015	0.5982	0.0495	0.6492
N_2O	(g)	0.0009	0.0104	0.0011	0.0124
SOx	(g)	0.1389	5.6483	0.1947	5.9819
NOx	(g)	0.1223	1.478	0.0739	1.674
CO	(g)	0.1638	0.2572	0.0236	0.4447
PM	(g)	0.0235	0.3411	0.0537	0.4183
VOC	(g)	0.0011	0.2224	0.00054	0.2240

Table 23: Greenhouse Gas Emissions from Transportation
(Per gal of FT Fuel Transported)

		Truck	Tanker	Pipeline	Total
Scenario 1, 3 & 4	(g CO_2-eq)	28.61	0	5.35	33.96
Scenario 2	(g CO_2-eq)	28.61	0	26.74	55.35
Scenario 5	(g CO_2-eq)	28.61	225.57	32.08	286.26
Scenario 6	(g CO_2-eq)	28.61	465.80	21.39	515.80

The most significant factors in determining transportation related emissions are fuel type and overall distance traveled (delivery and return trips). The combustion of RFO generates larger emissions of criteria pollutants than DFO and electricity generation and tanker routes are longer.

Fugitive emissions for intermediate product storage (marine and distribution terminals) along the various routes are expected to be insignificant relative to transportation and distribution and have been ignored for the LCI.

6. FUEL COMBUSTION, EFFICIENCIES & ANCILLARY EMISSIONS

This section contains a summary of ancillary emissions used in this LCI to estimate emissions along the FT fuel chain, and other factors required for estimating full life-cycle emission on a per vehicle mile basis.

6.1 Emissions Inventory for Ancillary Feedstocks

Emission factors for ancillary feedstocks were compiled from a number of sources [6,20,21,22] and are given in Table 24. The feedstocks of interest are electricity used in mining, FT production and pipeline transportation of FT products; low-sulfur, distillate fuel oil (DFO) used for tank truck distribution of FT products; high-sulfur, distillate fuel oil used by surface mining equipment; residual fuel oil (RFO) used in tanker transportation of FT products; fuel gas used in FT production, and butanes and methanol used to upgrade FT products. Upstream emissions are included in these factors, except for fuel gas, which is generated at the FT plant. Electricity emissions are based on a standard mix of power generation sources in the U.S. of 51% coal, 3% fuel oil, 15% natural gas, 20% nuclear, and 11% renewable sources.

Table 24: Emissions Inventory for Ancillary Feedstocks

	Electricity	Diesel Truck	Heavy Equip.	Tanker	Fuel Gas	Butane	Methanol
	Delivered	Delivered & Consumed	Delivered & Consumed	Delivered & Consumed	Consumed	Delivered	Delivered
	(g/kWh)	(g/MM Btu)	(g/MM Btu)	(g/MM Btu)	(g/MM Btu)	(g/bbl)	(g/bbl)
MM Btu/bbl	-	5.83	5.83	6.29	-	-	-
CO_2	711	80503	80503	86680	calculated	25859	11172
CH_4	1.76	4.3	4.3	15.2	1.3	92	112
N_2O	0.042	2.6	2.0	2.0	2.0	0.84	1.59
SOx	6.92	396	454	1088	0.0	8.1	102
NOx	1.8	348	937	818	63.6	149	165
CO	0.205	466	404	303	15.4	34.7	37.8
VOC	1.81	93.2	68.4	152	2.7	215	225
PM	1.91	66.9	70.53	97.50	1.36	6.7	11.1

6.2   Combustion Properties of Selected Fuels

Table 25 lists the CO_2 emissions factors for full combustion of the various products from the FT plant designs described in Section 3. These values are used to estimate the carbon emissions for end-use combustion of FT fuels. Also given in Table 25 are the emissions associated with the flaring and venting of associated gas; these are used in the sensitivity analysis presented in Section 7.3.

Table 25: CO_2 Emissions from Combustion of Selected Fuels

FT Gasoline/Naphtha	Wt. % C	g CO_2/gal
Design Option 1	85.63	8551
Design Option 2	85.05	8408
Design Option 3	78.73	7825
Design Option 4	85.63	8550
Design Option 5	86.81	8813
Design Option 6	85.95	8602
Design Options 7, 8	84.60	8058
FT Distillate		
Design Options 1, 2, 4-8	84.60	9011
Design Option 3	84.86	8956
	Wt. % C	g CO_2/Mscf
Flared Associated Gas	61.96	55984
	Wt. % C	g CO_2-eq/Mscf
Vented Associated Gas	61.96	313521

6.3   Vehicle Fuel Economies

The case study and sensitivity analysis presented in Sections 7.2 and 7.3 are for SUVs powered by conventional and advanced compression-ignition diesel engines. In order to estimate emissions for this study or others to be considered in the future, it is necessary to have an estimate of fuel economies for various vehicles and technologies. Table 26 contains fuel economies in units of miles-per-gallon (mpg) for various existing and future vehicle technologies based on efficiency estimates prepared by Argonne National Laboratory (ANL) [23]. It assumes spark-ignition engines are currently fueled by petroleum-derived gasoline and compression-ignition engines are fueled by petroleum-derived diesel fuel. The hybrid engine technologies consider on-board electricity generation and storage, and are not considered in this LCI.

Given mpg for one vehicle and technology, an estimate for the same vehicle with a different technology can be estimated from Table 26. The values in this table are based on the average energy content of petroleum-derived gasoline and diesel used in the U.S. Since FT fuels will have different energy contents than those derived from petroleum, the fuel economies in Table 26 must be adjusted based on the ratio of the heating value of the FT fuel to heating value of the petroleum fuel. For FT diesel this factor is 0.92.

Table 26: Vehicle Fuel Economy-Technology Matrix*
(miles-per-gallon)

Spark Ignition									
Conventional	10.0	15.0	20.0	25.0	30.0	35.0	40.0	45.0	50.0
Hybrid Electric	16.3	24.4	32.5	40.6	48.8	56.9	65.0	73.1	81.3
Direct Injection	12.7	19.0	25.3	31.6	38.0	44.3	50.6	57.0	63.3
Hybrid/Direct Inject	19.2	28.8	38.5	48.1	57.7	67.3	76.9	86.5	96.2
Compression Ignition									
Conventional	13.3	20.0	26.6	33.3	40.0	46.6	53.3	59.9	66.6
Advanced	15.3	23.0	30.6	38.3	46.0	53.6	61.3	68.9	76.6
Hybrid Electric	20.0	30.1	40.1	50.1	60.1	70.2	80.2	90.2	100.2
Advanced Hybrid	23.1	34.6	46.1	57.6	69.2	80.7	92.2	103.7	115.3

*For FT fuel multiply mpg by 0.92.

Comparisons between vehicles powered by gasoline spark-ignition and diesel compression-ignition engines must be done carefully. While there is a clear relationship between fuel economy and engine type, the basis for the comparison must also include the same type of vehicle used in similar applications (i.e., city or highway driving). For example, the average fuel economy for gasoline-powered passenger cars in the U.S. is about 30.7 mpg, for gasoline-powered SUVs it is 20 mpg, and for light-duty diesel-powered vehicles it is about 39 mpg. In similar applications, diesel engines are 33% more efficient than gasoline engines (from Table 26, (13.3 - 10.0 mpg)/10.0 mpg = 0.33). Therefore, converting all SUVs powered by gasoline to diesel would result in a fuel economy increase from 20 to 26.6 mpg (not to 39 mpg). Fuel composition also plays an important role in fuel economy. Substituting FT diesel for petroleum diesel in today's diesel-powered vehicles would result in a decrease in fuel economy from about 39 to 35.8 mpg, an 8% decrease. This is a result of the inherent lower energy density per gallon of FT diesel relative to conventional petroleum diesel.

7. FULL FT-FUEL LIFE-CYCLE INVENTORY

Six baseline scenarios were identified for consideration in this study. They involve the evaluation of different options for the resource extraction, conversion, and transportation/distribution steps in the FT fuel chain. Descriptions of these scenarios are given below.

Scenario 1

Production of FT fuels from bituminous Illinois No. 6 coal at a mine-mouth location in southern Illinois. The mine is an underground longwall mine. The design of the FT conversion plant is based on Option 1 described in Section 3. Upgrading includes a full slate of refinery processes for upgrading FT naphtha. Hydrocracking is used to convert the FT wax into additional naphtha and distillate. The liquid fuel products are shipped by pipeline to a terminal in the Chicago area and distributed by tank truck to refueling stations in the immediate area.

Scenario 2

Production of FT fuels from subbituminous Powder River Basin coal at a mine-mouth location in Wyoming. The mine is a surface strip mine. The design of the FT conversion plant is based on Option 4 described in Section 3. Upgrading steps are identical to those used in Scenario 1. The liquid fuel products are shipped by pipeline to a terminal in the Chicago area and distributed by tank truck to service stations in the immediate area.

Scenario 3

Production of FT fuels from plantation biomass (maplewood chips) at a location in southern Illinois. The design of the FT conversion plant is based on Option 5 described in Section 3 and co-produces electric power. Some naphtha upgrading is included; however, no LPG product is produced. Hydrocracking is used for FT wax conversion. The liquid fuel products are shipped by pipeline to a terminal in the Chicago area and distributed by tank truck to service stations in the immediate area.

Scenario 4

Production of FT fuels from pipeline natural gas at a location in southern Illinois. The design of the FT conversion plant is based on Option 6 described in Section 3. Upgrading steps are identical to those used in Scenarios 1. The liquid fuel products are shipped by pipeline to a terminal in the Chicago area and distributed by tank truck to service stations in the immediate area.

Scenario 5

Production of FT fuels from associated natural gas (of same composition as ANS gas) at a wellhead location near the coast of Venezuela. The design of the FT conversion plant is based on Option 7 described in Section 3. FT wax hydrocracking is included; however, no upgrading of the naphtha is performed. The liquid fuel products are shipped by tanker to a U.S. Gulf Coast marine terminal. From there they are shipped by pipeline to a terminal in the Chicago area and distributed by tank truck to service stations in the immediate area.

Scenario 6

Production of FT fuels from associated natural gas at a wellhead location on the Alaska North Slope. The design of the FT conversion plant is based on Option 7 described in Section 3 and is identical to that used for Scenario 5. The liquid fuel products are shipped by the Trans-Alaskan Pipeline to Valdez on the southern coast of Alaska. There they are transferred to a tanker for shipment to a marine terminal in the San Francisco Bay area and distributed by tank truck to service stations in the immediate area.

7.1   Emissions Inventory for Full FT Fuel Chain

Table 27 contains the LCI for the six scenarios described in the preceding section. This was compiled from the individual inventories for the resource extraction, conversion, and transportation/distribution steps of the FT fuel chain described in Sections 3, 4 and 5 of this report. *They are the full inventories up through the point of sale of the FT fuel* and are based on the entire FT liquid-fuel product slate. That is, the individual products (LPG, gasoline/naphtha, and distillate fuel) have not been broken out separately. Refueling and end-use combustion are not included. Refueling emissions are related to the volatility of the fuel. Because FT distillate is composed primarily of high-boiling paraffins, the volatility of diesel fuel is very low, and refueling emissions can be neglected in the LCI. The volatility of FT naphtha or gasoline derived from this naphtha will depend on the upgrading of this stream, and fugitive emissions for this product are not considered further in this analysis. The inclusion of end-use combustion emissions, other than CO_2, in the inventory requires specification of the end-use combustion device and its efficiency. Section 7.2 considers GHG emissions for the specific application of FT diesel in diesel-powered SUVs. In general, the emissions from FT diesel combustion are low; however, further work will be necessary to characterize the CP emission reduction benefits of FT fuels for specific vehicle applications.

Table 27: Emissions Inventory for FT Fuels at Point of Sale
(Per gal of FT Fuel Supplied)

		Scenario 1	Scenario 2	Scenario 3	Scenario 4	Scenario 5	Scenario 6
CO_2	(g)	12850	13865	-6564	4236	6385	6607
CH_4	(g)	26.0	3.76	0.45	14.9	6.07	6.36
N_2O	(g)	0.0582	0.08	0.65	0.08	0.09	0.096
SOx	(g)	5.82	8.61	0.19	0.23	3.22	6.03
NOx	(g)	2.50	3.34	17.8	11.7	10.4	10.8
CO	(g)	0.57	0.68	5.33	2.98	2.46	2.49
VOC	(g)	1.71	2.47	2.66	16.5	13.2	13.2
PM	(g)	1.49	2.35	0.30	0.06	0.30	0.45

Emissions reported in Table 27 follow the trends observed in Table 4 for the FT production step. Most emissions are higher for the coal and biomass designs relative to the gas-to-liquid designs. FT production is the dominant source of all emissions upstream of end use combustion. The major

58

exception is CH_4 emissions from underground mining of Illinois No. 6 coal, which is the largest single source of CH_4 emissions in Scenario 1.

7.2   Case Study - Substitution of FT Diesel Fuel in SUVs

The results from the FT LCI were used to evaluate the substitution of FT diesel for petroleum-derived fuels in Sport Utility Vehicles (SUVs) and the effect this substitution would have on greenhouse gas emissions. SUVs are almost exclusively powered by conventional spark-ignition internal combustion engines and fueled with petroleum-derived gasoline. In the U.S. they average roughly 20 mpg. Mileage for SUVs could be significantly improved by the use of diesel compression-ignition engines, which are about 33% more efficient than gasoline spark-ignition engines. Their use would result in an improvement in fuel economy to about 26.6 mpg. However, conventional diesel engines are high emitters of criteria pollutants. It has been demonstrated that FT diesel produces emissions that are much lower than those from petroleum-derived diesel. There is, however, a penalty to fuel economy when using FT diesel due to its lower energy density per gallon relative to petroleum-derived diesel. FT diesel fuel economy in an SUV has been estimated to be about 24.4 mpg. The full fuel-chain GHG emissions inventory for Scenarios 1-6 is presented in Table 28.

Table 28: Full Life-Cycle GHG Emissions for FT Diesel
(g CO_2-eq/mile in SUV)

Scenario/ FT Plant Feedstock	Extraction/ Production	Conversion/ Refining	Transport./ Distribution	End Use Combustion	Total Fuel Chain
1) IL #6 Coal	26	543	1	368	939
2) Wyoming Coal	7	585	2	368	962
3) Plantation Biomass*	-969	703	1	368	104
4) Pipeline Natural Gas	71	121	1	368	562
5) Venezuelan Assoc. Gas	51	212	12	368	643
6) ANS Associated Gas	51	212	21	368	652

*-969 = -1011 absorbed by biomass + 42 emitted during production.

The end-use combustion emissions (368 g CO_2-eq/mile) have been assumed constant for all the scenarios. Minor differences in the diesel produced by the various FT plant designs have been ignored (only Option 3 produces a distillate with a significantly different carbon and energy content, and this design has not been selected for consideration in any of these scenarios). The scenarios analyzed all employed FT wax hydrocracking and, unlike petroleum-derived diesel, FT diesel is of consistent high quality, regardless of the feedstock used for its production.

The results presented in Table 28 illustrate a number of interesting points. Emissions from transportation (1 to 21 g CO_2-eq/mile) clearly correlate to the distance the FT fuel is moved to market. Transportation emissions are low (1 to 2 g/mile) for domestic coal and biomass based scenarios, due to the close vicinity of the coal field or plantation and the FT plant to the fuel market (Chicago). For the coal and biomass Scenarios 1, 2 and 3, the largest single source of emissions is the indirect liquefaction facility (543 to 703 g/mile), with GHG emissions even larger than those for end-use combustion. For pipeline natural gas, GTL conversion emissions (121 g/mile) are lower than those for end-use combustion. Carbon and oxygen must be removed from coal and biomass to convert them into a liquid. This step requires energy and consumes syngas. The GTL process extracts hydrogen from methane to produce liquid fuels. However, there is still a significant emissions penalty with GTL, due to the consumption of energy during conversion, with subsequent emissions of CO_2. If the produced natural gas contains significant quantities of CO_2, emissions of GHG from conversion can be dramatically higher, as can be seen by comparing Scenario 5 or 6 to Scenario 4 (212 vs. 121 g/mile, respectively).

While biomass conversion emissions are higher than those for coal (703 vs. 543-585 g CO_2-eq/mile); overall, the full-fuel chain GHG emissions for biomass-based FT fuels is very low (104 g/mile). Biomass is a renewable resource, and the carbon it contains is recycled between the atmosphere and the fuel (resulting in the fixation of 1011 g of atmospheric CO_2/mile in the biomass). However, biomass cultivation and harvesting result in GHG emissions (42 g/mile), and biofuels should not be considered CO_2 emissions free.

Table 29 contains the GHG emissions per kWh for electric power produced and sold by the FT plants in Scenarios 3, 4 and 6d (6d is described in section 7.3). Also given for comparison are life-cycle GHG emissions for the average electricity generated in the U.S. (based on the results in Table 24) for typical existing, new and advanced PC (Pulverized-Coal) power plants using Illinois No. 6 coal [16] and for a biomass-gasification combined-cycle power plant based on the BCL design [19]. The allocation procedure used for fuels and power affects the relative values reported in Tables 28 and 29 for these scenarios. It is clear that for all the co-production scenarios, the GHG emissions for power generation are substantially lower than the norm for operating power generation plants in the U.S. The efficiencies reported in Table 29 for power production are total plant electrical efficiencies, whereas, those reported for the different scenarios only consider the actual power-producing device (gas or steam turbine) within the FT plant.

Table 29: Full Life-Cycle GHG Emissions for Power Exported from FT Plants
(g CO_2-eq/kWh of Electric Power)

Scenario/ FT Plant Feedstock	All Upstream	Electricity Generation	Total Fuel Chain	Electric Efficiency
3) Plantation Biomass	-1138	828	-309	60%
4) Pipeline Natural Gas	142	244	386	35%
6d) ANS Associated Gas	59	109	168	60%
U.S. Average All Plants	77	682	759	-
U.S. Average PC Plants	51	995	1045	32%
NSPS PC Plant	46	917	963	35%
LEBS PC Plant	21	722	743	42%
Biomass Gasification Combined-Cycle	-853	890	37	37%

The negative value (-309 g CO_2-eq/kWh) reported in Table 29 for Scenario 3 implies that the allocation procedure used skews the benefits of renewable biomass toward power generation relative to FT fuels production. This is also true for the natural gas-based designs that co-produce power.

7.3   Sensitivity Analysis

To help identify possible GHG reduction strategies for FT fuels production, a number of sensitivity cases were considered for the baseline scenarios described above. These included the application of advanced diesel engine technologies; coalbed methane capture, sequestration of process CO_2 from FT production; sequestration of process and combustion CO_2 from FT production; co-production of fuels and power; co-processing of coal and biomass; co-processing of coal and coalbed methane; and capture and conversion of flared or vented associated natural gas. Sequestration involves the collection, concentration, transportation and storage of CO_2 to reduce GHG emissions. Co-production refers to the production of multiple products from the indirect liquefaction plant; in this case, both fuels and power. Co-processing refers to the production of FT fuels from multiple feedstocks; for example, coal with biomass. Results are given in Table 30.

Table 30: Life-Cycle Sensitivity Analysis for FT Diesel
(g CO_2-eq/mile in SUV)

Scenario/ Modification to Baselines	GHG Emissions Reduction		Total Fuel Chain	
			existing diesel engine	advanced diesel engine
1a) IL #6 coal baseline	-	-	939	816
1b) with seq. of process CO_2	449	48%	490	426
1c) with seq. of process & comb. CO_2	516	55%	423	368
1d) with co-prod. of fuels & power	304	32%	635	552
1e) with co-proc. of biomass	155	17%	783	682
1f) with coalbed CH_4 capture	22	2.3%	917	798
1g) with co-proc. of coalbed CH_4	234	25%	705	613
4a) Pipeline natural gas baseline	-	-	562	489
4b) with seq. of process CO_2	65	12%	497	432
4c) with seq. of process & comb. CO_2	120	22%	442	384
5a) Venezuelan assoc. gas baseline	-	-	643	559
5b) with flaring credit	578	90%	65	57
5c) with venting credit	3234	503%	-2592	-2255
6a) ANS associated gas baseline	-	-	652	567
6b) with seq. of process CO_2	94	14%	558	485
6c) with seq. of process & comb. CO_2	211	32%	441	383
6d) with co-prod. of fuels & power	119	18%	534	464

The GHG emission reductions reported in Table 30 were estimated from the detailed energy and material balances reported for the conceptual process designs. However, they are <u>only</u> possible maximums since they do not include any analysis (re-design) of the conceptual FT process they were based on. They assume 100% recovery of CO_2 and CH_4 by the processes that might be used for the capture of these gases and ignore any possible energy penalties due to these processes.

For the production of FT fuels from fossil feedstocks, carbon (CO_2) sequestration would have the greatest impact on GHG emissions reductions. The sensitivity analysis presented in Table 30 shows that it might be possible to reduce GHG emissions from coal liquefaction by 48% (939 to 490 g CO_2-eq/mile for Scenario 1) and from GTL by 12-14% (562 to 497 and 652 to 558 g/mile for Scenarios 4 and 6, respectively), by sequestering the high-purity CO_2 stream being produced from the FT conversion plant. In addition, a significant quantity of CO_2 is generated from FT plant fuel combustion. If oxygen were used for combustion, this CO_2 could also be captured as a concentrated stream and sequestered, resulting in 55%, 22% and 32% reductions in total fuel-chain GHG emissions for Scenarios 1, 4 and 6, respectively. Both of these options would likely result in significant parasitic energy and cost penalties for the FT conversion process. However, these might be minimized by the application of new and developing technologies. Using pure CO_2 as a diluent could mitigate materials problems resulting from oxygen-rich combustion in fired heaters, boilers and gas turbines, and advanced oxygen production technologies could have significant benefits.

Sequestration shows less benefit for natural gas than for coal conversion. This results from less CO_2 being generated in the syngas generation and FT conversion steps for GTL. The larger total reduction for Scenario 6c relative to 4c (32 vs. 22%) is a result of the capture and sequestration of the 13% CO_2 present in the associated gas feedstock. The GHG emissions from coal or natural gas conversion are almost the same (423 vs. 441 g CO_2-eq/mile for Scenarios 1c and 4c/6c, respectively), if vented CO_2 and CO_2 from combustion are sequestered. The only remaining GHG emissions from FT production are fugitive and ancillary emissions, which are small and may also be reduced. The emissions from the natural gas scenarios with sequestration are even slightly larger than those from the coal scenario with sequestration. This is due to the higher production/extraction and transportation/distribution emissions for the natural gas scenarios considered here.

Scenario 6d considers the co-production of FT fuels and power. This estimate is based on FT plant design Option 8. Design Options 7 and 8 are identical except that Option 7 is self-sufficient in power and produces no excess electrical power for sale; whereas, Option 8 generates excess power from unconverted syngas and other plant fuel gas streams. This "once-through" conversion approach results in a 56% reduction in emissions from FT production, and an 18% reduction in total GHG emissions (from 652 to 534 g CO_2-eq/mile) based on the allocation procedure employed for this study. These gains are achieved by eliminating the recycle and reforming of off-gas produced in the FT conversion process. Assuming an equivalent percentage reduction in the FT conversion step of Scenario 1 results in a 32% reduction in full fuel-chain GHG emissions for indirect coal liquefaction (from 939 to 635 g/mile). *A detailed analysis is required to determine if this large of a reduction could actually be possible for a coal-based co-production facility.*

Co-processing of other feedstocks with coal may also be a viable approach to reducing GHG emissions. Scenarios 1e and 1g indicate that emissions could be cut roughly 17 to 25% from the coal conversion scenario (from 939 to 705-783 g CO_2-eq/mile) by co-feeding 20% biomass to gasification or by producing half the fuel product from methane rather than coal. Both these situations have other merits. The quantity of biomass available from a single plantation is quite small relative to the coal available from a single mine. At present, substitution of renewable biomass is hampered by the diffuse nature of this resource and is limited to at most 20% (LHV-basis) of the feed to a typical FT plant (50,000 bpd). Integrating the conversion of coal and biomass in a single co-processing facility would improve the economics of biomass conversion through shared economies of scale.

As discussed in Section 4, substantial quantities of methane are found associated with coal seams. Capture of coalbed methane from the mined seam only provides a small reduction in GHG emissions (2.3% based on Scenario 1f). If this methane were converted to FT fuels, it would only increase production by about 300 bpd for a 50,000 bpd plant. However, in certain coal producing regions, large quantities of coalbed methane could be produced from unmineable seams. Production of CH_4 from these seams can be stimulated by injecting CO_2 into the seam. Thus, this option provides an opportunity to sequester CO_2 produced from the FT process.

Scenarios 5b and 5c show the effect of reducing gas flaring and venting. In some parts of the world, significant amounts of associated gas are flared, because there is no readily available market for this natural gas. In Scenario 5b, it is assumed that the gas being used to produce the FT fuels was previously being flared. When credit is taken for eliminating flaring, full fuel-chain emissions are cut drastically (from 643 to 65 g CO_2-eq/mile). The situation is even more dramatic if this gas was simply being vented (from 643 to –2592 g/mile), since methane is such a potent greenhouse gas. Venting of associated gas was not uncommon only a few decades ago. The elimination of flaring and venting could under future regulations result in "carbon-credits" which could be sold in any market-based approach to reducing GHG emissions worldwide.

The last column in Table 30 lists the corresponding GHG emissions for SUVs powered by advanced diesel engines achieving 28.1 mpg, when operated on FT diesel. The net result of this next-generation vehicle technology is an across the board 13% reduction in emissions per mile. In general, CP emissions from FT diesel combustion are lower than those from petroleum-derived diesel, making FT diesel an ideal alternative to petroleum-derived diesel in advanced engines.

7.4   Comparison of FT and Petroleum-Derived Diesel Fuels

It is interesting to compare the results from the LCI for FT diesel to those for petroleum-derived diesel. Literature data were used to make this comparison. The petroleum-derived diesel estimates listed in Table 31 are based on information given in an article published by T.J. McCann & Associate Ltd. [24]. While these results cannot be independently verified, they have been reported to be from detailed private-client studies. As such, they can be assumed to include sources of data on emissions that are difficult or impossible to estimate without the involvement of petroleum producers, transporters and refiners. Based on crude oil properties and location, this information was used to estimate emissions for ANS and Wyoming crude oils. The GHG emissions for the other crude oils listed in Table 31 are from the original source.

The fuel chain for petroleum is similar to that shown for FT fuels in Figure 1 of Section 2, the major difference being that petroleum crude oil may be transported long distances prior to being refined into finished products. Crude oil transportation and refined-product transportation and distribution have been combined in Table 31. Again, transportation is a modestly significant source of emissions when crude oil is transported long distances (e.g. 26 g CO_2-eq/mile for Arab Light). Thus, in a carbon-constrained world, it may not make environmental sense to move oil (or any other commodity) halfway around the world.

There are significant differences between the GHG emissions for transportation from the McCann analysis relative to the FT LCI estimated here (e.g., 8 g CO_2-eq/mile for transporting Wyoming crude vs. 2 g/mile for FT syncrude from Wyoming coal). No explanation of these differences is possible without details of the McCann inventory. However, it is possible that the private client information reveals larger emissions from real-world operations.

While combustion dominates total emissions for petroleum, other contributing sources are not insignificant. Conversion and refining emissions (74-143 g CO_2-eq/mile), the second largest contributor, vary with crude API gravity. The API gravity is inversely proportional to specific gravity. High API gravity (low specific gravity) crude oils are generally of higher quality than low API gravity crude oils, which are referred to as heavy crudes. Heavier crude oils require more upgrading and refining and produce less desirable by-products. Emissions associated with their end-use are also higher, reflecting the poorer quality of their products. While not evident from the crude oils listed, production/extraction emissions are also related to crude API gravity. Heavier oils require reservoir stimulation techniques (such as steam injection), which require significant expenditures of energy and produce additional GHG emissions. Arab Light crude oil is an exception to the rule. Its high emissions result from flaring and venting of associated gas, a potential feedstock for GTL.

Table 31: Full Life-Cycle GHG Emissions for Petroleum Diesel
(g CO_2-eq/mile in SUV)

Crude Oil (°API)	Extraction/ Production	Conversion/ Refining	Transport./ Distribution	End Use Combustion	Total Fuel Chain
Wyoming Sweet (40°)	23	74	8	363	468
Canadian Light	30	81	11	367	489
Brent North Sea (38°)	23	81	8	367	479
Arab Light (38°)	35	81	26	367	509
Alaska North Slope (26°)	28	101	14	378	522
Alberta Syncrude (22°)	32	104	10	370	516
Venezuelan Heavy Oil (24°)	32	108	13	382	534
Venezuelan Syncrude (15°)	32	143	10	390	574

Comparing Tables 28 and 31, the production of FT diesel from coal results in significantly higher GHG emissions than for petroleum-derived diesel (962-939 vs. 468-574 g CO_2-eq/mile). GTL technology can achieve GHG emissions levels between those for coal liquefaction and petroleum refining (562-652 g/mile), due to the higher hydrogen content of methane relative to petroleum (4 to 1 vs. ~2 to 1). In fact, for natural gas Scenario 4, the GHG emissions for FT diesel are lower than the emissions for Venezuelan syncrude (562 vs. 574 g/mile), which requires severe processing to make it suitable as a feedstock for refining. Sequestration of vented CO_2 and CO_2 from combustion (Scenarios 1c, 4c and 6c) may be able to reduce GHG emissions to levels below those for products from petroleum refining. If advanced diesel engines are considered, then Scenarios 1b, 4b and 6d may also achieve these low GHG emissions levels.

7.5   Strategies for Reducing GHG Emissions from the FT Fuel Chain

The GHG emission reduction strategies identified in Section 7.3 can be divided into two categories: upstream and end-use. *Upstream GHG reduction strategies* involve modifications to the indirect liquefaction process in order to remove and sequester CO_2 produced during conversion, co-produce fuels and power, substitute biomass feedstocks, or mitigate the direct venting and flaring of methane. *End-use GHG reduction strategies* involve improvements in the efficiency of the end-use fuel application. With improved fuel efficiency less fuel is consumed per mile and less fuel must be produced and transported. Examples include adoption of higher-efficiency conventional and advanced diesel engines for passenger transportation (as was considered above for SUVs) or radical changes to the vehicular power plant (such as adoption of fuel cell technology in vehicles). These changes may also impact the processing used to produce the fuel owing to changes in fuel characteristics that their adoption might involve. In the extreme, they could necessitate *fuel switching*, the substitution of a totally new or different fuel for a given engine application. This is the main argument for replacing gasoline-powered engines with diesel-powered engines in SUVs.

The GHG reduction scenarios outlined below consider combinations of upstream and end-use strategies identified in the sensitivity analysis to maximize reductions:

GHG Reduction Scenario 7

Production of FT fuels from domestic coal reserves at a mine-mouth location. Locally available biomass is co-processed by co-feeding 20% biomass (LHV-basis) with the coal to produce liquid fuels. Any coalbed methane emissions from the mine are captured and also co-fed to the FT plant. The FT plant design is based on once-through conversion of the syngas and co-production of fuels and electric power. A portion of the power is used in the FT plant, and a portion is directed to coal mining operations. The remainder is sold, possibly generating GHG emission reduction credits.

Emissions Estimate:	Basis (Scenario 1a)	939 g/CO$_2$-eq/mile
	Co-processing of biomass (1e)	-155
	Co-production of power (1d)	-304
	Coalbed methane capture (80% of 1f)	- 18
		462
	Adv. diesel engine (13% reduction)	× .87
		402

A potential reduction of 537 g/CO$_2$-eq/mile or 57%.

GHG Reduction Scenario 8

Production of FT fuels from domestic coal reserves at a mine-mouth location. Locally available biomass is co-processed by co-feeding 20% biomass (LHV-basis) with the coal to produce liquid fuels. Any coalbed methane emissions from the mine are captured and also co-fed to the FT plant. The FT plant design is based on recycle of the unconverted syngas to maximize the production of liquid fuels; however, some electric power is co-produced to satisfy the needs of the FT plant and coal mine. Emissions of greenhouse gases from the plant are minimized by sequestering CO_2 in aquifers or other formations. Oxygen is used for combustion, thus producing an additional concentrated CO_2 stream for sequestration. Oxygen required for gasification and combustion may be supplied by advanced oxygen separation technologies. CO_2 is used as a diluent during combustion to control furnace, boiler and turbine temperatures.

Emissions Estimate:	Basis (Scenario 1a)	939 g/CO_2-eq/mile
	Co-processing of biomass (1e)	-155
	Sequestration of process CO_2 (90% of 1b)	-404
	Sequestration of combustion CO_2 (80% of 1c-1b)	- 54
	Coalbed methane capture (80% of 1f)	- 18
		308
	Adv. diesel engine (13% reduction)	× .87
		268

A potential reduction of 671 g/CO_2-eq/mile or 71%.

GHG Reduction Scenario 9

Production of FT fuels from domestic coal reserves at a mine-mouth location. Any coalbed methane emissions from the mine are captured and co-fed to the FT plant, along with coalbed methane recovered from the surrounding region. Thus, a substantial fraction of the feed to the plant is methane and half the fuel product is produced from methane rather than coal. The FT plant design is based on recycle of the unconverted syngas to maximize the production of liquid fuels; however, some electric power is co-produced to satisfy the needs of the FT plant, coal mine and coalbed methane operations. Emissions of greenhouse gases from the plant are minimized by sequestering CO_2 in unmined coal seams, thus enhancing the recovery of coalbed methane. Oxygen is used for combustion, thus producing an additional concentrated CO_2 stream for sequestration. Oxygen required for gasification and combustion may be supplied by advanced oxygen separation technologies. CO_2 is used as a diluent during combustion to control furnace, boiler and turbine temperatures.

Emissions Estimate:	Basis (Scenario 1a)	939 g/CO_2-eq/mile
	Co-processing of coalbed methane with credit for gas transmission & processing (average of 4a-1a+.95×71)	-222
	Sequestration of process CO_2 (90% of average of 1b+4b)	-231
	Sequestration of combustion CO_2 (80% of average of (1c-1b)+(4c-4b))	- 98
	Coalbed methane capture (80% of average of 1f+0)	- 9
		379
	Adv. diesel engine (13% reduction)	× .87
		330

A potential reduction of 609 g/CO_2-eq/mile or 64%.

It is expected that with current technology, significant parasitic energy losses would result from sequestration and increased use of oxygen in the FT plant. For the above estimates, it was assumed that only 90% of the vented CO_2 could be captured and sequestered, 90% of the CO_2 from combustion could be captured (0.9 x 0.9 x 100% = ~80% captured and sequestered), and 80% of coalbed methane emissions from mining could be captured. It was further assumed that results from the biomass co-production Scenario 1e and the pipeline gas Scenario 4a could be used to estimate emissions for coal and biomass and coal and coalbed methane co-processing, respectively. Since utilizing coalbed methane will not require cross-country transportation and processing requirements are minimal, credit was given in this scenario for a 95% reduction in extraction/production emissions. The benefits of co-production are based on the natural gas co-production Scenario 6d. No credit has been taken for the sale of the power co-produced, even though, GHG emissions will be lower than those from a typical existing power plant.

The analysis given above only identifies what may be possible. While Scenario 8 shows the biggest GHG emissions reduction relative to the other Scenarios 7 and 9 (71% vs. 57 and 64%), too much uncertainty exists in these estimates to consider one scenario better than another. *Further in-depth analysis will be needed to accurately quantify the future scenarios developed above, and technology breakthroughs will be required in CO_2 sequestration, oxygen separation, and combustion technology to achieve these benefits.*

8.0 CONCLUSIONS & RECOMMENDATIONS

The results of the life-cycle inventory and sensitivity analysis presented in Section 7 raise a number of new questions:

- Can realistic processes be developed to reduce or eliminate GHG emissions from the production of FT fuels from fossil energy resources?
- What is the actual resource base available for co-processing coal and biomass, or coal and coalbed methane?
- How should emissions be allocated between co-produced fuels and power?
- Can GHG emissions reduction credits be realized by co-producing power, elimination of venting and flaring of natural gas/coalbed methane, etc.?
- What might these credits be worth in the future?
- What will the GHG emissions from petroleum refining look like in the future?
- What are the GHG emissions from other advanced vehicle technologies: advanced spark-ignition engines, fuel cells, hybrid-electric systems, etc.?
- How do CP emissions from FT production and end-use compare with existing systems?
- What about emissions of water and solid waste from the production of FT fuels?
- What are the future technology needs to realize these GHG reductions?
- What might this all cost?

In order to answer these questions, life-cycle emissions and economic issues will need to be further addressed. These issues are discussed in more detail below.

8.1 Life Cycle Assessment

Questions regarding the optimal allocation of emissions between co-produced fuels and power, and determination of GHG reduction credits were beyond the scope of this study. Answers will require the careful comparison of existing energy and fuel systems. The allocation procedure used here for scenarios involving the co-production of fuels and power is based on standard practice within the LCA community. However, it can in many cases result in as many problems with the analysis as it solves. *Decisions are always made between alternatives. A preferred approach, therefore, would be to consider avoided or incurred emissions* due to the net production or consumption of electric power relative to some other alternative for providing this power. If net power is consumed at the FT plant, then emissions incurred by offsite power generation are added to the FT plant emissions as was done here. If net power is produced at the FT plant, emissions avoided from offsite power generation are subtracted. Whether power production at the FT plant is beneficial or not then depends on the basis used for offsite power generation. *Details of such an approach should be pursued in any further investigations.*

A more complex variation of the allocation problem also arises when comparing FT fuels to petroleum-derived fuels, where not only may product qualities differ, but the finished product and by-product mix can be significantly different. It has been suggested [24] that the various by-products

69

from petroleum refining (petroleum coke, LPG, home-heating oil, etc.) be debited to the premium products (gasoline, jet and diesel fuel) based on the assumption that natural gas could be substituted for these other fuels, if they were never produced. This same procedure could be used with FT fuels. Although, these problems were not considered in this LCI, they need to be addressed in the future.

It can also be foolhardy to only consider GHG emissions and ignore all other airborne, waterborne, or solid emissions. *Improvements relative to GHG reductions may very well be offset by other effects on the environment or human health and well being.* A preliminary inventory of upstream emissions from FT fuel production has been included here. Completing this inventory will require consideration of the end-use application, which in addition to SUVs, could include other gasoline or diesel powered vehicles or equipment, or even future hybrid or fuel cell powered vehicles. Analysis of fuel switching scenarios like these *will require expansion of the emissions inventory to future petroleum production and refining systems to establish a basis upon which to make comparisons of benefits and drawbacks.*

8.2   Economic Issues

It is clear that many of the GHG emissions reduction options considered here would be expensive to implement. Current estimates for the cost of indirect liquefaction (Bechtel ILBD) correspond to a required selling price for the FT products of roughly $1.24 per gal (1998 dollars before taxes and marketing charges). This price is based on updates (by E^{2}S-NETL) to the conceptual designs developed in the early 1990s. However, there is reason to believe that rapid technology improvement in oxygen separation, coal gasification, and FT conversion could lower this price by as much as $0.20 per gal. This, coupled with the premium which FT diesel is likely to command, puts FT fuels in a near-competitive range with petroleum-derived gasoline and diesel. *There is a need to update the analysis used to determine the required selling price and FT product premium to reflect current and future trends in transportation fuels markets.*

Recent DOE estimates for the cost of sequestration technologies (other than forest sinks) are well over $100 per ton of carbon sequestered. The estimates for future technologies under development range anywhere from $5 to $100 per ton ($1.4 to $27 per ton of CO_2). The DOE carbon sequestration program has a goal of driving down the cost of sequestration to $10 per ton through aggressive technology development. While the CO_2 emissions from indirect coal liquefaction are high, the process has a significant advantage in that CO_2 can be removed from the process as a concentrated stream that could easily be sequestered. Based on these estimates then, the cost of sequestration of process CO_2 from indirect liquefaction is about $0.33 per gal based on $100 per ton (0.449 kg CO_2/mile × 24.4 mile/gal × 2.2 lb/kg × 1 ton/2000 lb × 27 $/ton) and $0.02 per gal based on the DOE target of $10 per ton. *The broad range of this potential added cost, and the possibility that it could wipe-out the significant cost reductions obtained over the last decade, make it paramount that efforts to reduce the cost of FT conversion be continued.*

In the immediate future, only limited supplies of low-cost biomass are available for alternative uses. E^{2}S-NETL estimates the required selling price of FT fuels derived from biomass range anywhere from $2.00 to $2.31 per gal, depending on the source of the biomass. *Unless these costs can be reduced and the biomass resource base expanded, this option is likely to only play an incremental, albeit potentially important, role in GHG reduction strategies* (e.g., in meeting international targets). However, conversion of biomass to FT diesel, with the addition of sequestration of the concentrated CO_2 stream co-produced, is the only strategy when compared with those reported here that has the promising potential to be used as a "CO_2 sponge" to reduce atmospheric GHG levels. This scenario has not been considered here, but deserves future attention.

The optimum coupling of all three technologies: sequestration, co-production, and co-processing, may be a very attractive GHG mitigation strategy to minimize both GHG emissions and their cost impact on indirect liquefaction. Thus, there is a pressing need to carefully examine in detail both the technology options for GHG emissions reduction and their cost impact on the FT product.

8.3   Concluding Remarks

A Life-Cycle Inventory of greenhouse-gas emissions from FT fuel production has been completed. This analysis has identified and quantified the significant sources of GHG emissions from the FT fuel chain. Emissions from the FT conversion step can be comparable to those from end-use combustion. At the present, GHG emissions from the FT fuel chain are greater than those from the existing petroleum-based fuel chain. Coal-based conversion is at a significant disadvantage relative to petroleum; whereas, natural gas conversion is only moderately worse than the best petroleum refining, but better than the production and refining of heavy crude oils. In order for FT technology to be accepted in a world that is becoming more-and-more conscious of the effects of burning fossil fuels, it will be necessary to identify strategies and technologies for reducing GHG and other emissions. This study has been able to identify a number of possible approaches, including carbon sequestration, co-production of fuels and power, and co-processing of coal and biomass or coal and coalbed methane. Improvements in vehicle technology will also benefit the FT fuel chain by increasing fuel economy and, thus, reducing emissions per mile.

This analysis has also confirmed the findings of other researchers that extraction and transportation-related GHG emissions are much less than the emissions associated with conversion and end-use combustion of the fuel. However, this is not to say that these emissions categories should not be included in any full or streamlined LCI. These emissions can still be quite large relative to those from other industries and their reduction represent a significant challenge for coal, oil and gas production companies. Any analyst working outside of these organizations faces major challenges in identifying and quantifying all sources of emissions. Access to actual field data is necessary to accurately determine the true levels of emissions. Significant uncertainties still exist and too much credibility should not be given to absolute values. Relevant differences should provide reliable guidance to policy decisions.

In order to evaluate the full potential of GHG reduction strategies for FT fuel production, all of the options considered here require better data and a more rigorous analysis beyond the scope of this study. Neither has a total view of the environmental benefits and deficiencies of FT fuels been realized in this analysis. A GHG emissions inventory has been completed, but only the first step has been taken toward developing a complete life-cycle inventory of all FT fuel chain impacts. Emissions of criteria pollutants have been identified for combustion sources along the fuel chain.

Further work will be necessary to estimate emissions from vehicles fueled by FT diesel and gasoline and to expand this inventory to all categories of multimedia emissions.

This life-cycle greenhouse-gas emissions inventory for Fischer-Tropsch fuels is only the first phase of a comprehensive assessment to characterize the impact, both short and long term, of FT fuel production on the environment and on human health and well-being. Future research will be focused on expanding the current emissions inventory to include a broader range of multimedia emissions of interest to NETL programs, and on performing life-cycle inventory and economic analyses corresponding to the new low-emission FT process designs identified here.

GLOSSARY OF PROCESS TERMINOLOGY

Acid Gas – a gas stream containing a large percentage of H_2S and/or CO_2.

Alkylation – a refining process used to convert light hydrocarbon gases into a quality gasoline blending component.

Amine Absorption System – a process for removing H_2S and/or CO_2 from a gas stream by means absorption of the acid gas in an amine solvent (e.g., MDEA) which is continuously recycled and regenerated.

Associated Gas – methane and other light hydrocarbon gases recovered from petroleum production operations.

Autothermal Reforming – a process for producing syngas from pure methane or natural gas which combines partial oxidation and steam reforming reactions to balance heating and cooling requirements in the integrated system.

Biomass – any hydrogen and carbon containing substance produced by living or very recently living organisms.

Bituminous Coal – a rank of coal typically found in the eastern U.S. which is generally of moderate to good quality for combustion or liquefaction.

Catalytic Reforming – a refining technology used to convert low-quality naphtha into high-quality gasoline by removing hydrogen from hydrocarbons to form unsaturated ringed-compounds called aromatics.

Claus Unit – a process for converting H_2S into elemental sulfur.

Coal Ash – the mineral matter contained in coal.

Coalbed Methane – methane released from coal mining operations.

Coal Cleaning – processes for removing coal ash from coal.

Coal Preparation – processes for preparing coal for utilization either via combustion or liquefaction, including cleaning, drying and grinding.

Coal Rank – a relative rating scale for of coals which is indicative of the age, carbon content, volatile matter and heating value of the coal.

Combined-Cycle Power Plant – a power plant which produces electric power from an integrated gas and steam turbine system.

Crude Oil – a naturally occurring hydrocarbon-based oil.

Cryogenic Separation – separation processes which rely on differences in the volatility of compounds at temperatures significantly below ambient conditions.

Dehydration/Compression – a process for removing both heavier hydrocarbons and water from a gas stream.

Diesel Fuel – blends of hydrocarbon components with carbon numbers generally in the range of 16 to 18 that meet specifications for use in diesel-cycle (compression ignition) engines.

Distillate – a feed or intermediate stream that can be processed into components suitable for blending into jet or diesel fuel.

Field Condensate – a liquid hydrocarbon mixture produced at the natural gas wellhead.

Fischer-Tropsch Synthesis – a catalytic process for converting synthesis gas into liquid hydrocarbons.

Flared Gas – any gas stream that is produced from production, transportation or refining and processing which is incinerated before being discharged.

Fluid Catalytic Cracking – a refining process which converts oils into gasoline and diesel blending components by catalytically cracking large hydrocarbon molecules into smaller molecules in the absence of hydrogen in a fluidized bed reactor.

Fly Slag – coal ash removed from the syngas produced by gasification processes as small particles.

Fractionation – any physical separation process, such as distillation or extraction, used to separate individual or subgroups of components from a mixture.

Fuel Oil – any oil suitable for combustion in a conventional or advanced boiler system.

Gas Conditioning – the recovery of hydrocarbon liquids from a gas stream to make the gas suitable for transportation and sale.

Gasification – a process for producing syngas from a solid feedstock, such as coal or biomass, by reaction with oxygen and/or steam.

Gasoline – blends of hydrocarbon components generally with carbon numbers in the range of 5 to 10 that meet specifications for use in gasoline-cycle (spark ignition) engines.

Gas Plant – a plant which combines processes for the separation and purification of gas streams such as natural gas.

Gas Sweetening – the removal of H_2S and/or CO_2 from a gas stream to make the gas suitable for transportation and sale.

Gas-To-Liquids (GTL) – a process for converting natural gas to liquid fuels, such as FT liquids or methanol.

Hydrocracking – a refining process which converts oils into gasoline and diesel blending components by catalytically cracking large hydrocarbon molecules into smaller molecules in the presence of hydrogen.

Hydrolysis – processes that react gas impurities with water to facilitate their removal.

Hydrotreating – a refining process used to improve the quality of naphtha and distillate streams by adding hydrogen to the components of the stream.

Indirect Liquefaction – any process for converting a hydrogen and carbon containing solid or gas feedstock into a liquid which employs an intermediate step involving synthesis gas.

Isomerization – a refining process which converts straight-chain molecules to branched molecules.

Jet Fuel – blends of hydrocarbon components with carbon numbers generally in the range of 10 to 16 that meet specifications for use in turbine engines.

Liquefaction – processes for converting a solid or a gas to a liquid, refers both to chemical and physical conversions.

Liquefied Natural Gas (LNG) – a natural gas stream which has been refrigerated and compressed to make it liquid.

Liquefied Petroleum Gas (LPG) – a mixture of hydrocarbons that are gases at ambient conditions and are stored as liquids under pressure. Used here to specifically refer to mixtures of propane and propylene and mixtures of butenes and butanes.

Longwall Mining – a coal mining technique that removes all the coal from a coal seam inducing controlled ground subsidence.

Methyl-Diethanol Amine (MDEA) – a solvent used to remove H_2S and/or CO_2 from a gas stream.

Methyl Tert-Butyl Ether (MTBE) – an oxygen containing blending component for gasoline.

Naphtha – a feed or intermediate stream that can be processed into components suitable for blending into gasoline.

Natural Gas – a naturally occurring mixture of hydrocarbon gases.

Natural Gas Liquids (NGL) – propane, butanes and heavier hydrocarbons recovered from natural gas.

Natural Gasoline – pentane and heavier hydrocarbons recovered from natural gas.

Petroleum – any naturally occurring hydrocarbon-based liquid, including crude oils.

Partial Oxidation (POX) – a process for producing syngas from hydrocarbons which uses oxygen gas (from air) to supply oxygen to the reaction.

Pressure Swing Absorption (PSA) – a process used to recover hydrogen from a gas stream that employs a solid absorbent and operates cyclically.

Recycle Gas – unconverted synthesis gas which is returned to the FT reactor for further conversion.

Refining – integrated processes used to convert a crude or synthetic crude oil into salable products such as gasoline, jet and diesel fuel.

Residual Oil – the heavy oil remaining after the lighter products are distilled from crude oil.

Saturate – a hydrocarbon molecule that contains all aliphatic bonds.

Shell Claus Offgas Treating (SCOT) – a process used to convert sulfur in the tail gas back into H_2S for recycle to the Claus unit.

Scrubbing – a process that contacts raw syngas with water to remove entrained fine particulates.

Sequestration – the capture, concentration and long-term storage of CO_2.

Slag – coal ash removed from coal during gasification in a molten state and subsequently cooled to form a solid.

Slurry Bubble Column Reactor – a three-phase reactor for contacting syngas with catalyst.

Sour Water – an aqueous stream containing dissolved H_2S and/or CO_2.

Steam Reforming – a process for producing syngas from hydrocarbons which uses steam to supply oxygen for the reaction.

Strip Mining – a surface coal mining technique that removes the overlying soil and rock to expose the coal seam.

Stripping – a process for removing H_2S and/or CO_2 from an aqueous stream by distillation, including the regeneration step of an amine absorption system.

Subbituminous Coal – a rank of coal typically found in the western U.S. which is generally of low to moderate quality for combustion or liquefaction.

Supercritical Extraction – a fractionation process that employs a supercritical solvent to facilitate the absorption and separation of one component from another.

Synthetic Crude Oil or *Syncrude* – an oil which has been manufactured from alternative feedstocks which has properties similar to crude oil.

Synthesis Gas or Syngas – a mixture of hydrogen and carbon monoxide that can be chemically converted to liquid fuels or chemicals.

Tail Gas – the gas leaving a Claus unit which contains trace impurities that must be removed before venting.

Tert-Amyl Methyl Ether (TAME) – an oxygen containing blending component for gasoline.

Vented Gas – any gas stream that is produced from production, transportation or refining and processing which is directly discharged to the atmosphere.

Water Gas Shift – the reaction and reverse reaction of CO and H_2O to form H_2 and CO_2.

ZSM-5 Upgrading – a Mobil proprietary process that converts naphtha and distillate into components suitable for gasoline blending.

REFERENCES

1. *A Technical Framework for Life-Cycle Assessment*, SETAC Publications, Pensacola, FL (1991).

2. *A Conceptual Framework for Life-Cycle Impact Assessment*, SETAC Publications, Pensacola, FL (1993).

3. *Life-Cycle Assessment Data Quality: A Conceptual Framework*, SETAC Publications, Pensacola, FL (1993).

4. *Guidelines for Life-Cycle Assessment: A "Code of Practice"*, SETAC Publication, Pensacola, FL (1993).

5. *ISO 14040, 1997 Standard; ISO 14041, 1997 Standard; ISO/CD 14042, 1999 Standard; ISO/DIS 14043, 1999 Standard; ISO/TR 14048, 1999 Standard; ISO/TR 14049, 1999 Standard*, International Organization for Standardization publications.

6. *Emissions of Greenhouse Gases in the United States*, Energy Information Administration, U.S. Department of Energy, DOE/EIA-0573 (2000).

7. *Baseline Design/Economics for Advanced Fischer-Tropsch Technology, Final Report*, DOE Contract No. DE-AC22-91PC90027 (1998).

8. *ASPEN Process Flowsheet Simulation Model of a Battelle-Based Gasification, Fischer-Tropsch Liquefaction and Combined-Cycle Power Plant*, Refining and End Use Study of Coal Liquids, Topical Report, DOE Contract No. DE-AC22-83PC91029 (1998).

9. *Updated Baseline Design/Economics of Indirect Gas Conversion*, Refining and End Use of Coal Liquids, Topical Report (Draft), DOE Project No. DE-AC22-93PC91029.

10. "Process Design-Illinois No. 6 Coal Case with Conventional Refining," *Baseline Design/Economics for Advanced Fischer-Tropsch Technology, Topical Report Volume I*, DOE Contract No. DE-AC22-91PC90027 (1994).

11. "Process Design-Illinois No. 6 Coal Case with ZSM-5 Upgrading," *Baseline Design/Economics for Advanced Fischer-Tropsch Technology, Topical Report Volume II*, DOE Contract No. DE-AC22-91PC90027 (1994).

12. "Process Design-Western Coal Case with Conventional Refining," *Baseline Design/Economics for Advanced Fischer-Tropsch Technology, Topical Report Volume III*, DOE Contract No. DE-AC22-91PC90027 (1994).

13. "Natural Gas Fischer-Tropsch Case Baseline," *Design/Economics for Advanced Fischer-Tropsch Technology, Topical Report Volume VI*, DOE Contract No. DE-AC22-91PC90027 (1996).

14. "Process Flowsheet (PFS) Models, Appendices-Part II (1 of 2)," *Baseline Design/Economics for Advanced Fischer-Tropsch Technology, Topical Report Volume IV*, DOE Contract No. DE-AC22-91PC90027 (1994).

15. "Process Flowsheet (PFS) Models, Appendices-Part II (2 of 2)," *Baseline Design/Economics for Advanced Fischer-Tropsch Technology, Topical Report Volume V*, DOE Contract No. DE-AC22-91PC90027 (1994).

16. Spath, P.L., and Mann, M.K, *Life Cycle Assessment of Coal-fired Power Production*, National Renewable Energy Laboratory, Golden, CO, TP-570-25119 (1999).

17. "Options to Baseline Design," *Direct Coal Liquefaction Baseline Design and System Analysis, Final Report on Baseline and Improved Baseline, Volume III*, DOE Contract No. DE-AC22-90PC89857 (March 1993).

18. "Methodology for Estimating Methane Emissions from Coal Mining," *Inventory of U.S. Greenhouse Gas Emissions and Sinks: 1990-1997, Annex D*, U.S. EPA (1998).

19. Mann, M.K., and Spath, P.L., *Life Cycle Assessment of a Biomass Gasification Combined-Cycle Power System*, National Renewable Energy Laboratory, Golden, CO, TP-430-23076 (1997).

20. DeLuchi, M., *Emissions of Greenhouse Gases from the Use of Transportation Fuels and Electricity, Volume 2, Appendixes A-S*, Center for Transportation Research, Argonne National Laboratory, U.S. Department of Energy (Nov. 1993).

21. *Petroleum, Industry of the Future, Energy and Environmental Profile of the U.S. Petroleum Industry*, U.S. Department of Energy, Office of Industrial Technologies (1998).

22. *Electric Power Annual, Volume I*, Energy Information Administration, U.S. Department of Energy (1999).

23. "Well-to-Wheel Efficiency Analysis Sees Direct-Hydrogen Fuel Cells, Advanced Diesel Hybrids Comparable," *Gas-to-Liquids News*, April,1999.

24. "Crude Oil Greenhouse Gas Life Cycle Analysis Helps Assign Values For CO_2 Emissions Trading," *Oil & Gas Journal*, Feb., 1999.

25. *Annual Energy Outlook, With Projections to 2020*, Energy Information Administration, U.S. Department of Energy, DOE/EIA-0383(99), (1998).

26. Perry, R, H, and Green D.W. (editors), *Perry's Chemical Engineers' Handbook, Seventh Edition* (1997).

APPENDIX A

Life-Cycle Greenhouse-Gas Emissions Inventory For Fischer-Tropsch Fuels

Example Calculations

TABLE OF CONTENTS

SECTION 1

INTRODUCTION

1. INTRODUCTION

Appendix A Objectives:
- ❑ Present the material and energy balance data from a conceptual process design developed for the DOE in the 1990s for coal liquefaction using Illinois #6 underground coal (Design Case 1 of 8)

- ❑ Present the emission data for all processes upstream and downstream of the FT conversion plant for Design Case 1. (i.e. ancillary emissions, end use combustion…)

- ❑ Present various assumptions and estimations made throughout the inventory analysis

- ❑ Present step-by-step sample calculations for Design Case 1 to illustrate the methods of estimating greenhouse gas emission data

A detailed analysis using <u>only</u> Design Case 1 of Scenario 1 (FT production from Illinois #6 coal for use in the Chicago area) is presented here. The same equations, assumptions, methodology, etc. can be applied to Scenarios 2 through 6. Most of the results for Scenarios 2 through 6 are also summarized with Scenario 1 throughout the Appendix.

Greenhouse Gases Considered:
- ❑ CO_2 (carbon dioxide) from syngas production, FT synthesis, fossil-fuel combustion along the life cycle, and venting from natural gas production.
- ❑ CH_4 (methane) from fugitive plant and pipeline emissions, incomplete combustion or incineration (gas flaring), and coalbed methane release.
- ❑ N_2O (nitrous oxide) from fuel combustion and cultivation of biomass.

Criteria Pollutants Considered
- ❑ CO (carbon monoxide)
- ❑ NOx (nitrogen oxides)
- ❑ SOx (sulfur oxides)
- ❑ VOC (Volatile Organic Compounds)
- ❑ PM (Particulate Matter)

SECTION 2

ANCILLARY EMISSIONS

2. ANCILLARY EMISSIONS

The ancillary feedstocks of interest for Design Case 1 of Scenario 1 (Illinois #6) are:
- ❑ Electricity for coal mining
- ❑ Electricity for FT production
- ❑ Electricity used for pipeline transportation of FT products
- ❑ Low sulfur distillate fuel oil (DFO) for tank truck distribution of FT products
- ❑ Fuel gas used in FT production
- ❑ Butanes for FT product upgrading
- ❑ High sulfur distillate fuel oil (RFO) for tanker transportation of FT products (not used in Scenario 1)

A. Electricity Emissions

Includes airborne emissions from extraction of the fossil fuel (upstream) and fuel combustion for power generation at the power plant (downstream).

STEP 1: Data Collection

Table A1: CO_2-Equivalent Emissions of Individual Greenhouse
Gases from Power Plants and Upstream Processes
(g CO_2-eq/kWh)
[20, pg. D-23], [22]

Electricity Source	Coal Boiler	Fuel Oil Boiler	NG Boiler	NG Turbine	Nuclear Power
Average Energy Mix	51%	3%	15%		20%
Upstream processes					
CH_4	65.7	7.9	16.3	16.3	2.7
N_2O	0.4	5.3	0.7	0.7	0.7
NMOCs	0.4	3.3	1.1	1.1	0
CO	0.3	1.5	0.4	0.4	0.1
NOx	5.9	20.6	21.9	21.9	4.6
CO_2	29.3	141.8	72.0	72.0	45.9
Power Plant					
CH_4	0.1	0.2	0.0	3.6	0.1
N_2O	16.3	10.0	9.8	9.8	3.3
NMOCs	0.1	0.3	0.1	0.3	0.1
CO	0.4	0.5	0.6	1.7	0.1
NOx	102.5	71.0	54.7	41.1	4.9
CO_2	1075.4	875.9	606.3	605.2	6.5
All non-CO_2 gases	119.5	82.0	65.2	56.4	8.4
CO_2	104.6	1017.7	678.3	677.2	52.4

Table A2: Global Warming Potential (GWP) Mass Equivalency Factors
(kg of Gas per kg of CO_2)
[20, pg. O-9]

Gas	Equivalency Factor
CO_2	1
CH_4	21
N_2O	310
CO	3
NOx	40
NMOCs	11

STEP 2: Use the CO_2-equivalent emissions (Table A1), including the Average Energy Mix, and the Mass Equivalency Factors (Table A2) to calculate the gas emissions on a g/kWh basis. Note that the emissions are allocated among the energy sources using the average energy mix.

Methane Example:

$$CH_4 CoalBoiler\ Upstream\ (gCH_4 / kWh) = \left(\frac{65.7\,gCO_2 - eq / kWh}{21\,gCO_2 - eq / gCH_4} \right) = 3.13 \qquad \text{(Eq 2.1)}$$

*Consider average energy mix to calculate the total methane emissions from upstream processes in electricity production.
$$CH_4 TotalUpstr\ eam\ (gCH_4 / kWh) = (0.51 \times 3.13) + (0.03 \times 0.38) + (0.15 \times 0.78) + (0.20 \times 0.13) \qquad \text{(Eq 2.2)}$$

$CH_4 TotalUpstream(gCH_4/kWh) = 1.75$

Use Eq 2.1 and Eq 2.2 to calculate the remaining upstream and downstream GHG emissions.

Table A3: Emissions of Individual Greenhouse Gases from
Power Plants and Upstream Processes
(g/kWh)

Electricity Source	Coal Boiler	Fuel Oil Boiler	NG Boiler	NG Turbine	Nuclear Power	Total w/Energy mix
Upstream processes						
CH_4	3.13	0.38	0.78	0.78	0.13	1.75
N_2O	0.0013	0.0171	0.0023	0.0023	0.0023	0.002
NMOCs	0.0364	0.300	0.100	0.100	0.00	0.043
CO	0.100	0.500	0.133	0.133	0.033	0.094
NOx	0.1475	0.515	0.548	0.548	0.115	0.198
SOx						0.0
CO_2	29.30	141.8	72.0	72.0	45.9	39.5
VOCs (NMOCs+CH_4)						37.26
Power Plants						
CH_4	0.005	0.010	0.000	0.171	0.005	0.004
N_2O	0.05	0.030	0.030	0.030	0.010	0.03
NMOCs	0.01	0.030	0.010	0.030	0.010	0.010
CO	0.133	0.167	0.200	0.567	0.033	0.111
NOx	2.56	1.775	1.368	1.028	0.123	1.60
SOx						Calculated
CO_2	1075	876	606	605	7	671
VOCs(NMOCs+CH_4)						0.014

STEP 3: Calculate the upstream and downstream SOx and PM emissions from power plants using a different data source. (Emission data was not available from reference [20])

SOx Electricity Emissions:
SOx upstream = na   (Assume 0)
SOx combustion (lb/MMBtu) = 1.45   [21, pg. 16]
Electricity efficiency (Btu/kWh) = 10,500   [21, pg. 16]

$$SOxTotal\,(g/kWh) = \left(\frac{1.45\,lbs\,SOx}{1e6\,Btu}\right) \times \left(\frac{10,500\,Btu}{kWh}\right) \times \left(\frac{454\,g}{lb}\right) = 6.91 \qquad \text{(Eq 2.3)}$$

PM Electricity Emissions:
PM upstream = na   (Assume 0)
PM combustion (lb/MMBtu) = 0.4   [21, pg. 16]
Electricity efficiency (Btu/kWh) = 10,500   [21, pg. 16]

$$PMTotal\,(g/kWh) = \left(\frac{0.4\,lbs\,PM}{1e6\,Btu}\right) \times \left(\frac{10,500\,Btu}{kWh}\right) \times \left(\frac{454\,g}{lb}\right) = 1.91 \qquad \text{(Eq 2.4)}$$

Table A4: Total Ancillary Emissions from Electricity Production (Extraction + Combustion)

Gas	g/kWh
CO_2	710.10
CH_4	1.76
N_2O	0.042
CO	0.205
NOx	1.80
SOx	6.9
VOC	1.81
PM	1.91

B. Distillate Fuel Oil (DFO) Emissions for Light Trucks

Distillate fuel oil is considered to be a low sulfur diesel fuel used for transporting FT fuels from the tank farm (Chicago) to local refueling stations (60-mile radius). The total distillate fuel oil emissions consist of DFO production (refining) emissions and combustion emissions. CH_4, N_2O, NOx, CO and VOC distillate fuel emission data were available in reference [20], otherwise CO_2, SOx and PM are calculated via other sources.

STEP 1: Data Collection. CH_4, N_2O, NOx, CO and VOC distillate fuel oil emissions below include the production and combustion of distillate fuel oil. For example, 4.3 g of methane is emitted per 1 million Btu distillate fuel oil used by light trucks for transportation.

CH_4 (g/MM Btu) = 4.3 or (0.00947 lb/MM Btu)   [20, pg. A-10]
N_2O (g/MM Btu) = 2.6 or (0.00573 lb/MM Btu)   [20, pg. A-10]
NOx (g/MM Btu) = 348 or (0.767 lb/MM Btu)   [20, pg. A-10]
CO (g/MM Btu) = 466 or (1.028 lb/MM Btu)   [20, pg. A-10]

*VOC (g/MM Btu) = 93 or (0.2053 lb/MM Btu) [20, pg. A-10]
*Includes CH_4 and NMHCs

STEP 2: Calculate the CO_2 emissions

*Carbon = 19.95 MM tonne/Quadrillion Btu [6, pg. 30]
*At Full Combustion

$$DistCO_2(lb/MMBtu) = \left(\frac{19.95e6TonneC}{1e15Btu}\right) \times \left(\frac{2204.6lb}{Tonne}\right) \times \left(\frac{lbmolCO_2}{12.01lbCarbon}\right) \times \left(\frac{44lbCO_2}{lbmolCO_2}\right) = 161 \qquad \text{(Eq 2.5)}$$

Assumption: Since only combustion emissions were available, the amount was increased by 10% to account for upstream emissions.

CO_2 Distillate Fuel Oil (g/MM Btu) = 80503 or (177.4 lb/MM Btu)

STEP 3: Calculate SOx emissions. This includes SOx from distillate production, combustion and refinery sulfur plant.

<u>SOx from combustion</u> = 72.64 g/MM Btu or (0.160 lb/MM Btu) [21, pg. 16]

 Assumption: The SOx emissions from this reference is from <u>off-highway</u> diesel fuel, therefore only 20% of the total SOx combustion emissions will be considered since highway distillate fuels have lower sulfur specifications (~500 ppm).

<u>SOx from distillate fuel oil production (refining):</u>

 Total refinery SOx (MM lb/year) = 2001   [21, pg.16]
 Distillate fuel (MM bbl/year) = 126.7   [21, pg. 9]
 Total refined products (MM bbl/year) = 657.7   [21, pg. 9]
 Distillate fuel oil (MM Btu/bbl) = 5.825   [25]
 Residual fuel oil (MM Btu/bbl) = 6.287   [25]

$$\%refineryDFO = \left(\frac{126.7MMBblDist/Year}{657.7MMBblTotal/Year}\right) \times 100 = 19.26 \qquad \text{(Eq 2.6)}$$

Next, use this percentage and allocate the total SOx (2001 MM lb/year) to the distillate fuel oil pool.

$$SOxrefinery(lbSOx/Bbl) = \left(\frac{2001MMlbsTotalSOx}{Year}\right) \times \left(\frac{Year}{126.7MMBbl}\right) \times .1926 = 3.04 \qquad \text{(Eq 2.7)}$$

$$SOxrefinery(gSOx/MMBtu) = \left(\frac{3.04lbsSOx}{bbl}\right) \times \left(\frac{454g}{lb}\right) \times \left(\frac{bblDistillate}{5.825MMBtu}\right) = 276.2 \qquad \text{(Eq 2.8)}$$

<u>SOx from sulfur plant:</u>

 Sulfur production (ton/day) = 26,466 or (9,660,090 ton/year)   [21, pg. 5]
 *SOx = 91.56 lb SO_2/tons sulfur produced   [21, pg. 113]
 *From SCOT process and incinerator exhaust
 *Assume SO_2 = SOx

Determine the total SOx produced from the sulfur plant per year.

$$\text{(Eq 2.9)}$$

$$SOx(lb/Year) = \left(\frac{9660090tonSulfur}{Year}\right) \times \left(\frac{91.56lbSOx}{tonSulfur}\right) = 8.8e8$$

Next, use the percentage of distillate (19.26%) and allocate total SOx produced per year to the distillate fuel oil pool.

$$SOxDistillate(lbSOx/bbl) = \left(\frac{8.8e8lbSOx}{Year}\right) \times 0.19 \times \left(\frac{Year}{126.7 MMbblDistillateProduced}\right) = 1.34 \qquad \text{(Eq 2.10)}$$

$$SOxDistillate(gSOx/MMBtu) = \left(\frac{1.34lbSOx}{bblDistillate}\right) \times \left(\frac{454g}{lb}\right) \times \left(\frac{bblDistillate}{5.825MMBtu}\right) = 104.8 \qquad \text{(Eq 2.11)}$$

Total SO$_2$ distillate fuel oil emissions (Light Trucks):
Total (gSO$_2$/MM Btu) = refining emissions + sulfur plant + end use combustion
Total (gSO$_2$/MMBtu) = 276.2 + 104.8 + 72.6(0.20) = 395.5
Total (lbO$_2$/MMBtu) = 0.8711

STEP 4: Calculate the PM emissions from diesel end use combustion and production (refining) of distillate fuel oil using equations 2.6, 2.7, and 2.8.

PM from combustion:
PM combustion = 4.54 g/MM Btu or (0.01 lb/MM Btu) [21, pg. 16]
Assumption: The PM emissions in this reference is from <u>off-highway</u> diesel fuel, therefore only 20% of the total PM combustion emissions will be considered since highway distillate fuels have lower PM specifications.

PM from distillate fuel oil production (refining):
Total PM (MM lb/year) = 557		[21, pg.16]
Distillate fuel (MM bbl/year)= 126.7		[21, pg. 9]
Total refined products (bbl/year) = 657.7		[21, pg. 9]
Distillate fuel oil (MM Btu/bbl) = 5.825		[25]
Residual fuel oil (MM Btu/bbl) = 6.287		[25]

Use equations 2.7 and 2.8 to calculate the PM emissions from refining.
PM Refining (g/MM Btu) = 66.0

Total PM distillate fuel oil emissions (Light Trucks):
Total (g/MM Btu) = 66.8 or (0.1472 lb/MM Btu)

Table A5: Total Ancillary Emissions from
Distillate Fuel Oil (Light Trucks)
(Delivery + Consumption)

Gas	g/MM Btu
CO$_2$	80503
CH$_4$	4.3
N$_2$O	2.6
CO	466.4
NOx	348.3
SOx	395.5
VOC	93.2
PM	66.8

C. Distillate Fuel Oil Emissions for Heavy Equipment

This is assumed to be high sulfur diesel fuel used in heavy (off-highway) equipment for coal mining, etc. These values include emissions from distillate fuel production and combustion. The "Off-Highway" data from source [20, pg. A10] is used.

STEP 1: Data Collection

CH_4 (g/MM Btu) = 4.3 or (0.00947 lb/MM Btu)	[20, pg. A-10]
N_2O (g/MM Btu) = 2.0 or (0.004405 lb/MM Btu)	[20, pg. A-10]
CO (g/MM Btu) = 404.1 or (0.890 lb/MM Btu)	[20, pg. A-10]
NOx (g/MM Btu) = 936.5 or (2.063 lb/MM Btu)	[20, pg. A-10]
*VOC (g/MM Btu) = 68.4 or (0.15066 lb/MM Btu)	[20, pg. A-10]

*Includes CH_4 and NMHCs

STEP 2: Calculate the CO_2 emissions from distillate fuel production and combustion for <u>heavy equipment</u>.

> **Assumption:** CO_2 emissions are the same for heavy equipment as those calculated above for light trucks.
> *Same emission value as in step 2 of the distillate fuel (light truck) section.

> CO_2 Distillate Fuel Oil (g/MM Btu) = 80503 or (177.4 lb/MM Btu)

STEP 3: Calculate SOx emissions. This includes SOx from distillate fuel production, combustion and refinery sulfur plant for <u>heavy equipment</u> use.

> **Assumption:** SOx emissions from distillate fuel production and refinery sulfur plant is the same as for light trucks. Since off-highway has a higher sulfur specification (~5000 ppm), <u>total</u> combustion credit will be taken instead of using only 20% as for the light trucks.

> Total SOx Distillate Fuel Oil (g/MM Btu) = 453.63 or (1.0 lb/MM Btu)

STEP 4: Calculate the PM emissions from delivery and consumption of distillate fuel (<u>heavy equipment</u>) using equations 2.6, 2.7, and 2.8 above.

> **Assumption:** The same PM emissions will be generated for distillate fuel oil used by light trucks and heavy equipment <u>except</u> for combustion. The full PM value for combustion will be taken into consideration for the heavy equipment, but otherwise the same upstream production PM emissions are assumed to be equal.

> Total PM Distillate Fuel Oil (g/MM Btu) = 70.54 or (0.1554 lb/MM Btu)

Table A6: Total Ancillary Emissions from
Distillate Fuel Oil (Heavy Equipment)
(Delivery + Consumption)

Gas	g/MM Btu
CO_2	80503
CH_4	4.3
N_2O	2.0
CO	404.1
NOx	936.5
SOx	453.6
VOC	68.4
PM	70.53

D. Residual Fuel Oil (RFO) Emissions:

This is assumed to be the high sulfur diesel (off-highway) used for the tanker shipment of FT diesel fuel. Although a tanker is not used in Scenario 1, the calculations are shown here.

STEP 1: Data Collection

CH_4 (g/MM Btu) = 15.2 or (0.03348 lb/MM Btu)　　　　　　　　　[20, pg. A-10]

N_2O (g/MM Btu) = 2.0 or (0.004405 lb/MM Btu)　　　　　　　　[20, pg. A-10]

CO (g/MM Btu) = 303.0 or (0.6674 lb/MM Btu)　　　　　　　　　　[20, pg. A-10]

NOx (g/MM Btu) = 818.2 or (1.8022 lb/MM Btu)　　　　　　　　　[20, pg. A-10]

*VOC (g/MM Btu) = 151.6 or (0.3339 lb/MM Btu)　　　　　　　　[20, pg. A-10]

*Includes CH_4 and NMHCs

STEP 2: Calculate the CO_2 emissions from residual fuel oil production and combustion for <u>tanker transportation</u>.

*Carbon = 21.49 MM tonne/Quadrillion Btu　　　　　　　　　　　[6, pg. 30]

*At Full Combustion

$$RFOCO_2 (lb/MMBtu) = \left(\frac{21.49e6 TonneC}{1e15 Btu}\right) \times \left(\frac{2204.6 lb}{Tonne}\right) \times \left(\frac{lbmol CO_2}{12.01 lbCarbon}\right) \times \left(\frac{44 lbCO_2}{lbmol CO_2}\right) = 173.6$$　　(Eq 2.12)

Assumption: Since only combustion emissions were available, the amount was increased by 10% to account for upstream emissions.

CO_2 Residual Fuel Oil (g/MM Btu) = 86680 or (190.9 lb/MM Btu)

STEP 3: Calculate the SOx emissions. Use equations 2.6 to 2.11 and same methodology as used for the distillate fuel oil in light trucks.

<u>SOx from RFO combustion:</u>

SOx combustion (g/MM Btu) = 771.8 or (1.70 lb/MM Btu)　　　　　　[21, pg. 16]

SOx from RFO production (refining):

Total SOx (MM lb/year) = 2001	[21, pg16]
Residual Fuel (MM bbl/year) = 45.9	[21, pg. 9]
Total Refined Products (MM bbl/year) = 657.7	[21, pg. 9]
Distillate Fuel Oil (MM Btu/bbl) = 5.825	[25]
Residual Fuel Oil (MM Btu/bbl) = 6.287	[25]

Use equations 2.6, 2.7 and 2.8 to calculate the SOx in the residual fuel oil.

SOx RFO Production (g/MM Btu) = 219.7 or (0.48392 lb/MM Btu)

SOx from Sulfur Plant:

Sulfur (ton/day) = 26,466 or 9,660,090 ton S produced/year	[21, pg. 5]
*SOx = 91.56 lb SO_2/tons Sulfur produced	[21, pg. 113]
*From SCOT process and Incinerator Exhaust	
*Assume SO_2 = SOx	

SOx Sulfur Plant (g/MM Btu) = 97.13 or (0.2139 lb/MM Btu)

Total residual fuel oil SO_2 Emissions:
Total SO_2 residual fuel oil (g/MM Btu) = 1088.1 or (2.396 lb/MM Btu)

STEP 4: Calculate the PM emissions from delivery and consumption of residual fuel oil using equations 2.6, 2.7, and 2.8.

PM combustion:
PM combustion (g/MM Btu) = 36.32 or (0.080 lb/MM Btu)   [21, pg. 16]

PM from residual fuel oil production (refining):

Total PM (MM lb/year) = 557	[21, pg16]
Residual fuel (MM bbl/year) = 45.9	[21, pg. 9]
Total refined products (MM bbl/year) = 657.7	[21, pg. 9]
Distillate fuel oil (MM Btu/bbl) = 5.825	[25]
Residual fuel oil (MM Btu/bbl) = 6.287	[25]

Use equations 2.6, 2.7 and 2.8 to calculate the PM emissions from residual fuel oil production/refining.

PM RFO Production (g/MM Btu) = 61.17

Total PM emissions from residual fuel oil:
PM Total = PM RFO Combustion + PM RFO Production
PM Total = 97.5 g/MM Btu or (0.21476 lb/MM Btu)

Table A7: Total Ancillary Emissions from Residual Fuel Oil
(Delivery + Consumption)

Gas	g/MM Btu
CO_2	86680
CH_4	15.2
N_2O	2.0
CO	303
NOx	818.2
SOx	1088
VOC	151.6
PM	97.5

E. Fuel Gas Ancillary Emissions

This is the fuel gas <u>consumed</u> in the FT plant. Does not consider production.

STEP 1: Data collection.

CO_2 (g/MM Btu) = 56,029 or (123.4 lb/MM Btu)	[20, pg. A-10]
CH_4 (g/MM Btu) = 1.3 or (0.000286 lb/MM Btu)	[20, pg. A-10]
N_2O (g/MM Btu) = 2.0 or (0.0044 lb/MM Btu)	[20, pg. A-10]
CO (g/MM Btu) = 15.4 or (0.035 lb/MM Btu)	[20, pg. A-10]
NOx (g/MM Btu) = 63.6 or (0.1400 lb/MM Btu)	[21, pg. 16]
VOC (g/MM Btu) = 2.7 or (0.004 lb/MM Btu)	[21, pg. 16]
SOx (g/MM Btu) = 0.00	[21, pg. 16]
PM (g/MM Btu) = 1.36 or (0.003 lb/MM Btu)	[21, pg. 16]

Table A8: Total Ancillary Emissions from Fuel Gas Consumption

Gas	g/MM Btu
CO_2	56,029
CH_4	1.3
N_2O	2.0
CO	15.4
NOx	63.6
SOx	0.0
VOC	2.7
PM	1.36

F. Butane Emissions

Butane is produced from natural gas; therefore the emissions are based off the <u>associated</u> natural gas emissions.

STEP 1: Obtain NG <u>upstream production</u> pipeline emissions.

> **Assumption:** Natural gas <u>extraction</u> emissions are the same for butane production as for electricity generation. Convert natural gas pipeline emissions (Table A3) from kWh to MM Btu by using an efficiency conversion factor of 11314 Btu/kWh (as per reference).

Table A9: Natural Gas Pipeline Emissions

Gas	g/kWh	g/MM Btu
CH_4	0.78	69
N_2O	0.0023	0.20
CO	0.133	11.8
NOx	0.548	48.4
SOx	0.002	0.212
CO_2	72.0	6364
VOCs	.8762	77
PM	0	0

STEP 2: Calculate the <u>Associated Natural Gas (ANG)</u>:

> **Assumption:** CO_2, N_2O, CO, NOx, SOx, VOC, and PM associated natural gas emissions are 69.6% of the pipeline natural gas and CH_4 is 33.3 % of the pipeline natural gas.

Table A10: Associated Natural Gas (ANG) Emissions

Gas	g/MM Btu
CO_2	4427
CH_4	22.8
N_2O	0.146
CO	8.2
NOx	33.7
SOx	.147
VOC	53.6
PM	0

STEP 3: Calculate the emissions associated with the butane transportation.

> The associated natural gas emissions will be combined with the butane transportation emissions (Table A5-light trucks).

Data:
Butane (MM Btu/Bbl) = 4.023 [26]
Butane density (lb/gal) = 5.007 [26]
Kansas to So. Illinois (miles) = 500
Trucking Energy Consumption (Btu/ton-mile) = 1900 [20, pg. E-9]
CO_2 Distillate Fuel Oil (g/MM Btu) = 80503 or (177.4 lb/MM Btu)

Transportation:

$$Emissions(g/galBut) = (EnergyConsumed) \times (Dist.) \times (Dens.) \times (Emissions) \times (Conv.Fact)$$ (Eq 2.13)

Example: Carbon dioxide emission from butane transportation.

$$TruckCO_2(g/galBut) = (1900) \times (500) \times (1/2000) \times (5.007) \times (177.4/1e6) \times (454) = 191$$ (Eq 2.14)

Table A11: Butane Transportation Emissions

Kansas to Southern Illinois (500 miles)

Gas	g/gal Butane delivered
CO_2	191
CH_4	.012
N_2O	.0062
CO	.03898
NOx	.15117
SOx	.17276
VOC	.00216
PM	.01080

STEP 4: Combine emissions from butane production via associated natural gas (Table A10) and butane transportation emissions from Kansas to Southern Illinois.

Example: Total CO_2 emissions from butane production and delivery

$$Total(gCO_2/bbl) = \left(\left(\frac{191gCO_2}{gal} \right) \times \left(\frac{42gal}{bbl} \right) \right) + \left(\left(\frac{4427gCO_2}{MMBtu} \right) \times \left(\frac{4.023MMBtu}{bbl} \right) \right) = 25859$$ (Eq 2.15)

**Table A12: Total Ancillary Emissions from
Butane Production and Delivery**

Gas	g/bbl Butane delivered
CO_2	25859
CH_4	92
N_2O	0.84
CO	34.7
NOx	141.8
SOx	8.1
VOC	215
PM	6.7

G. Ancillary Emissions Summary

*Does not include methanol emissions since they are not used in Scenario 1. Same as Table 24 in main report.

Table A13: Emissions Inventory for Ancillary Feedstocks

	Electricity	Diesel Truck	Heavy Equip.	Tanker	Fuel Gas	Butane
	Delivered	Delivered & Consumed	Delivered & Consumed	Delivered & Consumed	Consumed	Delivered
	(g/kWh)	(g/MM Btu)	(g/MM Btu)	(g/MM Btu)	(g/MM Btu)	(g/bbl)
MM Btu/bbl	-	5.83	5.83	6.29	-	5.023
CO_2	710.54	80503	80503	86680	calculated	25859
CH_4	1.756	4.3	4.3	15.2	1.3	92
N_2O	0.0421	2.6	2	2	2.0	0.84
SOx	6.92	395.5	453.63	1088	0.0	8.1
NOx	1.8	348.3	936.5	818.2	63.6	141.8
CO	0.205	466.4	404.1	303	15.4	34.7
VOC	1.81	93.2	68.4	151.6	2.7	215
PM	1.91	66.9	70.53	97.49	1.36	6.7

SECTION 3

FISCHER-TROPSCH
PROCESS

3. FISCHER-TROPSCH PROCESS

A. Resource Consumption & Yields for FT Production

Material and energy balance data from the eight indirect liquefaction baseline designs (ILBD) developed by Bechtel (see main report) were used to generate the resource consumption and yield data for each FT scenario studied. The ILBD data is summarized in Table 2 of the main report. This baseline design data provides the groundwork required to inventory the GHG emissions for the FT conversion process.

FT Product Basis—1bbl of FT C3+ liquid product contains:
- C_3/C_4 LPG
- Gasoline/Naphtha
- Distillate

STEP 1: Data collection. Obtained from the Indirect Liquefaction Baseline Design study done by Bechtel [7].

Table A14: Design Case 1 of Scenario 1 Fischer-Tropsch Material Balance Input Data
[7]

	Ton/day	Bbl/day
Raw Materials		
Illinois #6 Coal:	18575	
Catalyst & Chemicals:	342	
Products		
LPG:	171	1922
Butanes:	-317	-3110
Gasoline/Naphtha:	3021	23943
Distillate:	3343	24686
Other Out Flows		
Slag:	2244	
Sulfur:	560	
CO_2 Removal:	28444	
CO_2 Gasifier Carrier Gas:	-3715	
S-Plant Flue Gas:	1086	
Utilities		
Electric Power (MW):	54	
Raw Water Make-Up (MM Gal/day):	14.46	

STEP 2: Calculate the resource consumption per barrel of FT liquid product. Recall that the liquid FT product includes C3/C4 LPG, gasoline/naphtha, and distillate.

$$Coal\ (ton\ /\ bblFT\ ) = \frac{18575\ ton\ /\ day}{(1922 + 23943 + 24686\ )bbl\ /\ day} = 0.36745 \qquad \text{(Eq 3.1)}$$

$$Butanes\ (bbl\ /\ bblFT\ ) = \frac{-(-3110)}{1922 + 23943 + 24686} = 0.062 \qquad \text{(Eq 3.2)}$$

$$Cat\ \&\ Chem(lb\ /\ bblFT) = \frac{342 \times 2000}{1922 + 23943 + 24686} = 13.52 \qquad \text{(Eq 3.3)}$$

$$RawWater\ (gal\ /\ bblFT) = \frac{14.46e6}{1922 + 23943 + 24686} = 286 \qquad \text{(Eq 3.4)}$$

$$Power\ (kWh\ /\ bblFT) = \frac{54 \times 1000 \times 24}{1922 + 23943 + 24686} = 25.79 \qquad \text{(Eq 3.5)}$$

STEP 3: Calculate the volume yield of each product per barrel of total FT liquid product.

$$C_3\ /\ C_4 (bbl\ /\ bblFT\ ) = \frac{1922\ bbl\ /\ day}{(1922 + 23943 + 24686\ )bblFT\ /\ day} = 0.038 \qquad \text{(Eq 3.6)}$$

$$Gas\ /\ Nap\ (bbl\ /\ bblFT\ ) = \frac{23943}{1922 + 23943 + 24686} = 0.474 \qquad \text{(Eq 3.7)}$$

$$Distillate\ (bbl\ /\ bblFT\ ) = \frac{24686}{1922 + 23943 + 24686} = 0.488 \qquad \text{(Eq 3.8)}$$

STEP 4: Calculate the mass yield per barrel of FT liquid product.

$$C_3\ /\ C_4\ (ton\ /\ bblFT\ ) = \frac{171}{1922 + 23943 + 24686} = 0.003 \qquad \text{(Eq 3.9)}$$

$$Gas\ /\ Nap\ (ton\ /\ bblFT\ ) = \frac{3021}{1922 + 23943 + 24686} = 0.062 \qquad \text{(Eq 3.10)}$$

$$Distillate\ (ton\ /\ bblFT\ ) = \frac{3343}{1922 + 23943 + 24686} = 0.066 \qquad \text{(Eq 3.11)}$$

$$Slag\ (ton\ /\ bblFT\ ) = \frac{2244}{1922 + 23943 + 24686} = 0.044 \qquad \text{(Eq 3.12)}$$

$$Sulfur\ (ton\ /\ bblFT\ ) = \frac{560}{1922 + 23943 + 24686} = 0.011 \qquad \text{(Eq 3.13)}$$

STEP 5: Use the lower heating values of to calculate the energy yield per barrel of FT liquid product.

Table A15: Lower heating values (LHV)
[7]

	M Btu/lb
Coal:	11.95
Butanes:	19.6
LPG:	19.9
Gasoline/Naphtha:	17.7
Distillate:	18.9

$$Coal_{LHV}(MMBtu/day) = \left(\frac{18575\,ton\,coal}{day}\right) \times \left(\frac{2000\,lb}{ton}\right) \times \left(\frac{119500\,Btu}{lb}\right) \times \left(\frac{1}{1e6}\right) = 443744 \qquad \text{(Eq 3.14)}$$

$$C_3/C_{4_{LHV}}(MMBtu/day) = \frac{171 \times 2000 \times 19.9}{1000} = 6816 \qquad \text{(Eq 3.15)}$$

$$Butane_{LHV}(MMBtu/day) = \frac{-317 \times 2000 \times 19.6}{1000} = -12448 \qquad \text{(Eq 3.16)}$$

$$Gas/Nap_{LHV}(MMBtu/day) = \frac{3021 \times 2000 \times 17.7}{1000} = 107185 \qquad \text{(Eq 3.17)}$$

$$Distillate_{LHV}(MMBtu/day) = \frac{3343 \times 2000 \times 18.9}{1000} = 126365 \qquad \text{(Eq 3.18)}$$

*Divide the energy content of each product by the total FT liquid product.

$$C_3/C_4(MMBtu/bblFT) = \frac{6816}{(1922 + 23943 + 24686)} = 0.135 \qquad \text{(Eq 3.19)}$$

$$Gas/Nap(MMBtu/bblFT) = \frac{107185}{(1922 + 23943 + 24686)} = 2.12 \qquad \text{(Eq 3.20)}$$

$$Distillate(MMBtu/bblFT) = \frac{126365}{(1922 + 23943 + 24686)} = 2.50 \qquad \text{(Eq 3.21)}$$

STEP 6: Calculate the thermal efficiency per barrel of FT liquid product.

$$TotalFT_{LHV} = LPG_{LHV} + Butane_{LHV} + Gas/Nap_{LHV} + Distillate_{LHV} \qquad \text{(Eq 3.22)}$$

$$TotalFT_{LHV}(MMBtu/day) = 6816 - 12448 + 107185 + 126365 = 227919 \qquad \text{(Eq 3.23)}$$

FT process power required (used):

$$Power_{LHV}(MMBtu/day) = \frac{54\,MW \times 24\,hr/day}{0.2930711\,MW/MMBtu/hr} = 4449 \qquad \text{(Eq 3.24)}$$

$$\text{(Eq 3.25)}$$
$$ThermalEfficiency(\%) = \frac{TotalFT_{LHV} - Power_{LHV}}{Coal_{LHV}}$$

$$OptionThermalEffciency(\%) = \frac{227919 - 4449}{443744} \times 100 = 50.4\%$$

STEP 7: Calculate the carbon efficiency per barrel of FT liquid product.
The carbon efficiency for each case is calculated from the carbon balance data around the FT plant.

Case 1 carbon efficiency (coal):

Carbon out = 5292.8 ton/day [7]

Carbon in = 13190.1 ton/day [7]

$$CarbonEff. \ (\%) = \frac{Cout}{Cin} \times 100 \qquad \text{(Eq 3.26)}$$

Case 1 carbon efficiency (%) = 40.1

*This method is used to determine the carbon efficiencies for Design Cases 2, 3 & 4.

Case 5 carbon efficiency (biomass):

$$CarbonIN \ (ton/bblFT) = \left(\frac{0.621 \, tonBiomass \ Feed}{bblFT} \right) \times \left(\frac{0.49 \, tonC \ *}{tonBiomass} \right) = 0.3043 \qquad \text{(Eq 3.27)}$$

*0.49tonC/tonBiomass is from Table A36: Ultimate Analysis

$$CarbonOUT \ (ton/bblFT) = \left(\frac{131 \, ton}{day} \right) \times \left(\frac{day}{157 \, bblFT} \right) = 0.1132 \qquad \text{(Eq 3.28)}$$

Case 5 carbon efficiency (%) = 37.2

Case 6 carbon efficiency (pipeline gas):

$$CarbonIN \ (ton/bblFT) = \left(\frac{8.927 \, Mscf}{bblFT} \right) \times \left(\frac{8949 \, tonNG \ /day}{412000 \ Mscf \ /day} \right) = 0.193902 \qquad \text{(Eq 3.29)}$$

$$CarbonOUT \ (ton/bblFT) = \left(\frac{4971 \, ton}{day} \right) \times \left(\frac{day}{44602 \, bblFT} \right) = 0.111452 \qquad \text{(Eq 3.30)}$$

Case 6 carbon efficiency (%) = 57.5

*Design Cases 7 and 8 (Associated NG) use the same method and equations as for Design Case 6.

Table A16: Resource Consumption and Yields for FT Production
(Per bbl of FT Liquid Product)

	Case 1	Case 2	Case 3	Case 4	Case 5[1]	Case 6[1]	Case 7	Case 8[1]
Feedstock	IL #6	IL #6	IL #6	Wyo. Coal	Biomass	Pipeline Gas	Assoc. Gas	Assoc. Gas
Upgrading	Maximum Distillate	Increased Gasoline	Maximum Gaso. & Chem.	Maximum Distillate	Fuels & Power	Maximum Distillate	Minimum Upgrading	Min. Upgrading & Power
Resources								
Coal or Biomass (MF ton)	0.3675	0.3661	0.3310	0.395	0.621 [0.00072]			
Natural Gas (Mscf)						8.927 [0.018]	10.305	10.325 [0.012]
Butanes (bbl)	0.062		0.093	0.062		0.008		
Methanol (bbl)			0.041					
Catalysts & Chemicals (lb)	13.52	15.44	na	15.71	na	0.13	na	na
Water Make-Up (gal)	286	285	279	196	541 [0.629]	455 [0.923]	114	91 [0.105]
Electric Power (kWh)[2]	25.79	24.87	24.87	42.12	-1781	-13.2		-230
Volume Yield (bbl)								
C3/C4 LPG	0.038	0.071	0.118	0.038		0.038		
Gasoline/Naphtha	0.474	0.616	0.708	0.474	0.330	0.379	0.313	0.312
Distillates	0.488	0.313	0.174	0.488	0.670	0.583	0.687	0.688
Mass Yield (ton)								
C3/C4 LPG	0.003	0.007	0.011	0.003		0.003		
Gasoline/Naphtha	0.060	0.077	0.089	0.060	0.042	0.048	0.038	0.038
Distillates	0.066	0.043	0.023	0.066	0.091	0.079	0.092	0.092
Slag (MF)	0.044	0.044	0.040	0.035	0.065 [0.000075]			
Sulfur	0.011	0.011	0.010	0.002				
Energy Yield (MMBtu)								
C3/C4 LPG	0.135	0.262	0.422	0.134		0.134		
Gasoline/Naphtha	2.120	2.764	3.019	2.121	1.463	1.687	1.439	1.433
Distillates	2.500	1.611	0.862	2.498	3.427	2.979	3.495	3.494
Power[3]					10.128	0.128		1.309
Allocation to Fuels					0.326	0.974		0.790
Carbon Efficiency (%)	40.1	41.1	37.7	39.1	37.2	57.0	39.3	39.2
Thermal Efficiency (LHV)	50.4%	52.0%	47.4%	49.3%	51.0%	59.1%	57.3%	57.1%

1   Values in [ ] are allocations per kWh of electricity produced and sold. All other values are per bbl.

2   Positive value is purchase, negative value is sale.

3   Energy content of fuel used to produce power for sale.

Fischer Tropsch

B. Emissions Inventory for Fischer-Tropsch Production

STEP 1: Perform a carbon balance around the FT process to determine <u>all</u> GHG emissions.

*Note: Ultimate analysis data on FT feedstocks and FT products are contained in Table A36 at the end of Appendix A and is used throughout the following calculations.

$$C_{coal}(ton/day) = \frac{Coal_{Feed}\ ton/day \times \%Carbon}{100} \qquad \text{(Eq 3.31)}$$

$$C_{coal}(ton/day) = \frac{18575 ton/day \times 71.01}{100} = 13190 \qquad \text{(Eq 3.32)}$$

$$C_{LPG}(ton/day) = \frac{171 \times 81.72}{100} = 139.7 \qquad \text{(Eq 3.33)}$$

$$C_{Butanes}(ton/day) = \frac{-317 \times 82.66}{100} = -262 \qquad \text{(Eq 3.34)}$$

$$C_{Gas/Nap}(ton/day) = \frac{3021 \times 85.63}{100} = 2586.9 \qquad \text{(Eq 3.35)}$$

$$C_{Distillate}(ton/day) = \frac{3343 \times 84.6}{100} = 2828.2 \qquad \text{(Eq 3.36)}$$

$$C_{Slag}(ton/day) = \frac{2244 \times 3.36}{100} = 75.4 \qquad \text{(Eq 3.37)}$$

$$C_{CO_2 Vented}(ton/day) = \frac{(CO_2 Removed - CO_2 CarrierGas) \times \%C_{CO_2}}{100} \qquad \text{(Eq 3.38)}$$

$$C_{CO_2 Vented}(ton/day) = \frac{(28444 + (-3715)) \times 27.29}{100} = 6749 \qquad \text{(Eq 3.39)}$$

$$C_{CO_2 Misc}(ton/day) = \frac{0.01 \times 28444 \times 27.29}{100} = 77.6 \qquad \text{(Eq 3.40)}$$

$$C_{S-PlantFlu}(ton/day) = \frac{1086 \times 24.93}{100} = 270.7 \qquad \text{(Eq 3.41)}$$

The remaining carbon is from fuel gas combustion.

$$C_{GasCombuston}(ton/day) = C_{Coal} - C_{FTLTotal} - C_{Slag} - C_{CO_2 Vented} - C_{CO_2 Misc.} - C_{S-Plant} = 725 \qquad \text{(Eq 3.42)}$$

Table A17: Carbon Balance around FT Plant
(Design Case 1-Illinois #6 Coal)

Feedstock	Carbon (ton/day)
IL #6 Coal	13190
Energy Products	
LPG	139.7
Butanes	-262.0
Gasoline/Naphtha	2586.9
Distillates	2828.2
Total FTL	5292.8
Other Outflows	
Slag	75.4
Balance of Carbon	7821.9
CO_2 Vented (net removed)	6748.5
CO_2 Misc. Emissions	77.6
S-Plant Flue Gas	270.7
Fuel Gas Combustion	725.1

STEP 2: Combine the carbon balance data (Table A17) and ancillary emissions data (Table A13) to determine the FT process GHG emissions

<u>Carbon Dioxide Emissions:</u>

   CO_2 sources:

1.	Venting	5.	Power
2.	Misc. sources	6.	Butane
3.	Sulfur Plant		
4.	Fuel gas Combustion		

$$CO_{2\,Vented}\ (g/day) = \left(\frac{44.01}{12.01}\right) \times 6749\ tonC \times \left(\frac{453.6\,g}{lb}\right) \times \left(\frac{2000\ lb}{ton}\right) = 2.24\,e10 \tag{Eq 3.43}$$

$$CO_{2\,Fugitive}\ (g/day) = \left(\frac{44.01}{12.01}\right) \times 77.6\,tonC \times \left(\frac{453.6\,g}{lb}\right) \times \left(\frac{2000\,lb}{ton}\right) = 2.58\,e8 \tag{Eq 3.44}$$

$$CO_{2\,S-Plant}\ (g/day) = \left(\frac{44.01}{12.01}\right) \times 270.7\,tonC \times \left(\frac{453.6\,g}{lb}\right) \times \left(\frac{2000\ lb}{ton}\right) = 8.99\,e8 \tag{Eq 3.45}$$

$$CO_{2\,FuelGas}\ (g/day) = \left(\frac{44.01}{12.01}\right) \times 725.1\,tonC \times \left(\frac{453.6\,g}{lb}\right) \times \left(\frac{2000\ lb}{ton}\right) = 2.41\,e9 \tag{Eq 3.46}$$

$$CO_{2\,Power}\ (g/day) = 710.54\,g/kWh \times 54.32e6W \times 1000 \times \left(\frac{24\,hr}{day}\right) = 9.26\,e8 \tag{Eq 3.47}$$
*Ancillary CO_2 for power = 710.54 g/kWh

$$CO_{2\,Butanes}\ (g/day) = -(42\,gal/bbl) \times (615.69\,g/gal\,Butane) \times (-3110\,bbl\,Butane/day) = 8.04\,e7 \tag{Eq 3.48}$$

CO_2 Total (g/day) = 2.7e10

$$CO_2\,(g/bbl\,FTProduced) = \left(\frac{2.70e10}{1922 + 23943 + 24686}\right) = 534311 \tag{Eq 3.49}$$

Methane Emissions:

CH_4 sources:

    1. FT Plant fugitive, tank, and flaring emissions
2. FT Plant fuel combustion
3. Power
4. Butanes

Data:

Fuel Consumption LHV (MM Btu/hr): 1125.5 or (27012 MMBtu/day)   [7]

Fuel gas HHV (M Btu/lb): 5.18   [7]

Fuel gas LHV (M Btu/lb): 4.74   [7]

CH_4 (fugitive, tanks, flaring)(g/day) = 349081   [7]

$$FuelConsumptionHHV\ (MMBtu/day) = 27012 \times \left(\frac{FuelHHV}{FuelLHV} \right) \qquad \text{(Eq 3.50)}$$

$$FuelConsumptionHHV\ (MMBtu/day) = 27012 \times \left(\frac{5.18}{4.74} \right) = 29519 \qquad \text{(Eq 3.51)}$$

$$CH_{4\,FuelCombustion}\ (g/day) = 27012 \times 1.3 = 35116 \qquad \text{(Eq 3.52)}$$
*Ancillary CH_4 for power = 1.3 g CH_4/MM Btu

$$CH_{4\,Power}\ (g/day) = 1.756gCH_4/kWh \times 54.3MW \times 1000 \times 24hrs/day = 2.29e6 \qquad \text{(Eq 3.53)}$$

$$CH_{4\,Butanes}\ (g/day) = 92gCH_4/bblButane \times 3110bbl/day = 286058 \qquad \text{(Eq 3.54)}$$

$$CH_4 Total(g/day) = 349081 + 35116 + 2.29e6 + 286058 = 2.96e6 \qquad \text{(Eq 3.55)}$$

$$CH_4(g/bblFT) = \left(\frac{2.96e6}{1922 + 23943 + 24686} \right) = 58.6 \qquad \text{(Eq 3.56)}$$

Nitrous Oxide Emissions:

N_2O sources:

    1. FT Plant fuel gas combustion
2. Power
3. Butanes

$$N_2O_{FuelCombustion}(g/day) = 27012 \times 2.0 = 54024 \qquad \text{(Eq 3.57)}$$
*Ancillary N_2O = 2.0 gN2O/MM Btu Fuel Combusted

$$N_2O_{Power}\ (g/day) = .0421gN_2O/kWh \times 54.3MW \times 1000 \times 24hrs/day = 54890 \qquad \text{(Eq 3.58)}$$

$$N_2O_{Butanes}(g/day) = (0.84g/bblButane) \times (-3110bbl/day) = 129.6 \qquad \text{(Eq 3.59)}$$

$$N_2O Total(g/day) = 54024 + 54890 + 129.6 \cong 109043 \qquad \text{(Eq 3.60)}$$

$$N_2O(g/bblFT) = \left(\frac{109043}{1922 + 23943 + 24686} \right) = 2.16 \qquad \text{(Eq 3.61)}$$

Sulfur Oxides Emissions:

SOx sources:

1. Flue gas incineration
2. Power
3. Butanes

$$SOx_{FlueGas}(g/day) = FlueGasFlowrate \times \%Sulfur_{FlueGas} \tag{Eq 3.62}$$

$$SOx_{FlueGas}(g/day) = \left(\frac{1086tonFG}{day}\right) \times \left(\frac{.0005095lbS}{lbFG}\right) \times \left(\frac{2000lb}{ton}\right) \times \left(\frac{64.066lbSOx/lbmolSOx}{34.08lbS/lbmolSOx}\right) \times \left(\frac{453.6g}{lb}\right) = 943349 \tag{Eq 3.63}$$

$$SOx_{Power}(g/day) = 6.92gSOx/kWh \times 54.3MW \times 1000 \times 24hr/day = 9022241 \tag{Eq 3.64}$$

$$SOx_{Butanes}(g/day) = 8.1g/bbl \times 3110bbl/day = 25222 \tag{Eq 3.65}$$

$$SOxTotal(g/day) = 943349 + 9022241 + 25222 = 9990812 \tag{Eq 3.66}$$

$$SOx(g/bblFT) = \left(\frac{9990812}{1922 + 23943 + 24686}\right) = 197.6 \tag{Eq 3.67}$$

Nitrogen Oxides Emissions:

NOx sources:

1. Fuel gas combustion
2. Power
3. Butanes

$$NOx_{FuelCombustion}(g/day) = 63.6g/bbl \times 27012bbl/day = 1715370 \tag{Eq 3.68}$$

$$NOx_{Power}(g/day) = 1.8 \times 54.3MW \times 1000 \times 24hr/day = 2346826 \tag{Eq 3.69}$$

$$NOx_{Butanes}(g/day) = 141.54g/bbl \times 3110bbl/day = 441033 \tag{Eq 3.70}$$

$$NOxTotal(g/day) = 1715370 + 2346826 + 441033 = 4503229 \tag{Eq 3.71}$$

$$NOx(g/bblFT) = \left(\frac{4503229}{1922 + 23943 + 24686}\right) = 89.1 \tag{Eq 3.72}$$

Carbon Monoxide Emissions:

CO sources:
1. Fuel gas combustion
2. Power
3. Butanes

$$CO_{FuelCombustion}(g/day) = 15.4g/MMBtu \times 27012 MMBtu/day = 416590 \qquad \text{(Eq 3.73)}$$

$$CO_{Power}(g/day) = 0.205g/kWh \times 54.3MW \times 1000 \times 24hr/day = 267277 \qquad \text{(Eq 3.74)}$$

$$CO_{Butanes}(g/day) = 34.7g/bbl \times 3110bbl/day = 107802 \qquad \text{(Eq 3.75)}$$

$$COTotal(g/day) = 416590 + 267277 + 107802 = 791669 \qquad \text{(Eq 3.76)}$$

$$CO(g/bblFT) = \left(\frac{791669}{1922 + 23943 + 24686}\right) = 15.7 \qquad \text{(Eq 3.77)}$$

Volatile Organic Carbon Emissions:

VOC sources:
1. Fuel gas combustion
2. Power
3. Butanes

$$VOC_{FuelCombustion}(g/day) = 2.7g/MMBtu \times 27012 MMBtu/day = 73516 \qquad \text{(Eq 3.78)}$$

$$VOC_{Power}(g/day) = 1.81g/kWh \times 54.3MW \times 1000 \times 24hr/day = 2359864 \qquad \text{(Eq 3.79)}$$

$$VOC_{Butanes}(g/day) = 215g/bbl \times 3110bbl/day = 670511 \qquad \text{(Eq 3.80)}$$

$$VOCTotal(g/day) = 73516 + 2359864 + 670511 = 3103890 \qquad \text{(Eq 3.81)}$$

$$VOC(g/bblFT) = \left(\frac{3103890}{1922 + 23943 + 24686}\right) = 61.4 \qquad \text{(Eq 3.82)}$$

Particulate Matter Emissions:

PM sources:
1. Fuel gas combustion
2. Power
3. Butanes

$$PM_{FuelCombustion}(g/day) = 1.36g/MMBtu \times 27012 MMBtu/day = 36758 \qquad \text{(Eq 3.83)}$$

$$PM_{Power}(g/day) = 1.91 \times 54.3MW \times 1000 \times 24hr/day = 2490243 \qquad \text{(Eq 3.84)}$$

$$PM_{Butanes}(g/day) = 67/bbl \times 3110bbl/day = 20782 \qquad \text{(Eq 3.85)}$$

$$PMTotal(g/day) = 36758 + 2490243 + 20782 = 2547783 \qquad \text{(Eq 3.86)}$$

$$PM(g/bblFT) = \left(\frac{2547783}{1922 + 23943 + 24686}\right) = 50.4 \qquad \text{(Eq 3.87)}$$

Table A18: Emissions Inventory for FT Production
(Per bbl of FT Liquid Product)

	Case 1	Case 2	Case 3	Case 4	Case 5*	Case 6*	Case 7	Case 8*
Feedstock	IL #6	IL #6	IL #6	Wyo. Coal	Biomass	Pipeline Gas	Assoc. Gas	Assoc. Gas
Upgrading	Maximum Distillate	Increased Gasoline	Maximum Gaso. & Chem.	Maximum Distillate	Fuels & Power	Maximum Distillate	Minimum Upgrading	Min. Upgrading & Power
CO_2 (g)	534311	526684	507159	575203	706987	119687	210964	92978
CH_4 (g)	58.55	51.14	64.40	87.27	12.97	8.45	4.77	4.79
N_2O (g)	2.16	1.91	2.11	2.85	16.50	1.60	2.02	3.17
SO_x (g)	197.64	190.73	193.85	298.04	0	0.06	0	0
NO_x (g)	89.08	72.07	98.31	118.82	523.90	51.93	64.15	100.51
CO (g)	15.66	11.73	18.02	19.09	127.23	12.61	15.58	24.41
VOC (g)	61.40	46.19	76.21	91.05	22.45	3.77	2.75	4.31
PM (g)	50.40	48.10	49.53	81.60	11.23	1.14	1.37	2.15

* Values reported only include allocation to fuel products.

Fischer-Tropsch

A-36

C. Emissions Inventory for Power Exported from FT Plants

Design Cases 5, 6 and 8 produce significant excess power for sale. Therefore, it was necessary to allocate emissions between power and fuels in order to make comparisons with the other cases. The procedure used for this allocation has significant effect on the reported emissions per bbl of fuel produced. This uncertainty is compounded by a lack of information on fuel gas generation and consumption for some of the baseline designs. Therefore, caution should be exercised when comparing the emissions from biomass liquefaction to coal liquefaction, or emissions from the various natural gas cases. Example calculations for Design Case 5 will be presented here. Design Cases 6 and 8 follow the same method.

STEP 1: Calculate the energy yields for Design Cases 5, 6 and 8 using equations 3.15 through 3.21. See Table A16 "Resource Consumption and Yields for FT Production" above.

<u>Case 5 energy yields:</u>

Gas/Naphtha (MM Btu/bbl FT) =	1.463
Distillates (MM Btu/bbl FT) =	3.427
Power Sales (MM Btu/bbl FT) =	10.128
Power Sales (kWh/bbl FT) =	1781

<u>Case 5 FT process emissions (Table A18):</u>

CO_2 (g/bbl FT) =	706987
CH4 (g/bbl FT) =	12.97
N2O (g/bbl FT) =	16.50
SOx (g/bbl FT) =	0
NOx (g/bbl FT) =	523.9
CO (g/bbl FT) =	127.23
VOC (g/bbl FT) =	22.45
PM (g/bbl FT) =	11.23

STEP 2: Determine the allocation of power to fuels utilizing the HHVs and LHVs.

$$HHVFuelAllocation = \frac{(Gas/NapEnergy + Distillate\ Energy)}{(Gas/NapEnergy + Distillate\ Energy + PowerSales)} \quad \text{(Eq 3.88)}$$

$$HHVFuelAllocation = \frac{(1.463 + 3.427)}{(1.463 + 3.427 + 10.128)} = .326 \quad \text{(Eq 3.89)}$$

$$LHVFuelAllocation = 1 - HHVFuelAllocation = .674 \quad \text{(Eq 3.90)}$$

STEP 3: Calculate the emissions for <u>exported power</u> from FT plants. Use the component emissions from Table A18 and allocate them to power based on the HHV and LHV percentages.

$$CO_2(g/kWhPower) = \frac{(CO_2(g/bblFT) \times LHVAllocation)}{(PowerSales(kWh) \times HHVFuelAllocation)} \quad \text{(Eq 3.91)}$$

$$CO_2(g/kWhPower) = \frac{(706987g/bblFT \times 0.674)}{(1781kWh/bblFT \times 0.326)} = 822 \quad \text{(Eq 3.92)}$$

$$CH_4(g/kWhPower) = \frac{(12.97 \times 0.674)}{(1781 \times 0.326)} = 0.015 \qquad \text{(Eq 3.93)}$$

$$N_2O(g/kWhPower) = \frac{(16.50 \times 0.674)}{(1781 \times 0.326)} = 0.019 \qquad \text{(Eq 3.94)}$$

$$SO_x(g/kWhPower) = \frac{(0 \times 0.674)}{(1781 \times 0.326)} = 0.000 \qquad \text{(Eq 3.95)}$$

$$NO_x(g/kWhPower) = \frac{(523.9 \times 0.674)}{(1781 \times 0.326)} = 0.609 \qquad \text{(Eq 3.96)}$$

$$CO(g/kWhPower) = \frac{(127.23 \times 0.674)}{(1781 \times 0.326)} = 0.148 \qquad \text{(Eq 3.97)}$$

$$VOC(g/kWhPower) = \frac{(22.45 \times 0.674)}{(1781 \times 0.326)} = 0.026 \qquad \text{(Eq 3.98)}$$

$$PM(g/kWhPower) = \frac{(11.23 \times 0.674)}{(1781 \times 0.326)} = 0.013 \qquad \text{(Eq 3.99)}$$

Table A19: Emissions Inventory for Power Exported from FT Plants
(Per kWh of Electric Power)

	Case 5*	Case 6*	Case 8*
Feedstock	Biomass	Pipeline Gas	Assoc. Gas
Upgrading	Fuels & Power	Maximum Distillate	Min. Upgrading & Power
CO_2 (g)	822	243	107
CH_4 (g)	0.015	0.017	0.006
N_2O (g)	0.019	0.003	0.004
SOx (g)	0.000	0.000	0.000
NOx (g)	0.609	0.105	0.116
CO (g)	0.148	0.026	0.028
VOC (g)	0.026	0.008	0.005
PM (g)	0.013	0.002	0.002

*Values reported only include allocation to exported power.

D. Greenhouse Gas Emissions for FT Production

Greenhouse gas emissions for the FT designs have been compiled in Table A21. Emissions of CH_4 and N_2O have been converted to CO_2 equivalents using the global warming potentials (Table A20) for a 100-year time horizon.

Table A20: Global Warming Potentials for Selected Gases
(kg of Gas per kg of CO_2)
[6, pg. 8]

GAS	Lifetime (years)	Direct Effect for Time Horizons of:		
		20 Years	100 Years	500 Years
Carbon Dioxide (CO_2)	Variable	1	1	1
Methane (CH_4)	12 +/- 3	56	21	7
Nitrous Oxide (N_2O)	120	280	310	170

STEP 1: Use the GWPs and component emission data (Section B above) to calculate the GHG emissions from FT production on a per barrel FT basis.

$$CO_{2Vented}(g/bblFT) = \frac{22434556368}{(1922+23943+24686)} = 443800 \tag{Eq 3.100}$$

$$CO_{2FuelGasCombustion}(g/bblFT) = \frac{2410548571}{(1922+23943+24686)} = 47685 \tag{Eq 3.101}$$

$$CO_{2FlueGasIncineration}(g/bblFT) = \frac{899966486}{(1922+23943+24686)} = 17803 \tag{Eq 3.102}$$

$$CO_{2Fugitive}(g/bblFT) = \frac{258048044}{(1922+23943+24686)} = 5105 \tag{Eq 3.103}$$

$$CO_{2Ancillary}(g/bblFT) = \frac{926396368+80420775}{(1922+23943+24686)} = 19917 \tag{Eq 3.104}$$

$$CH_{4FuelCombustion}(gCO_2-eq/bblFT) = 21\times\frac{35116}{(1922+23943+24686)} = 15 \tag{Eq 3.105}$$

$$CH_{4Fugitive}(gCO_2-eq/bblFT) = 21\times\frac{349081}{(1922+23943+24686)} = 145 \tag{Eq 3.106}$$

$$CH_{4Ancillary}(gCO_2-eq/bblFT) = 21\times\frac{2289459+286220}{(1922+23943+24686)} = 1070 \tag{Eq 3.107}$$

$$N_2O_{Combustion}(gCO_2-eq/bblFT) = 310\times\frac{54024}{(1922+23943+24686)} = 331 \tag{Eq 3.108}$$

$$N_2O_{Ancillary}(gCO_2-eq/bblFT) = 310\times\frac{54890+130}{(1922+23943+24686)} = 337 \tag{Eq 3.109}$$

$$TOTAL\ (gCO_2-eq/bblFT) = 536209 \quad \text{*Sum of Eqs 4.100 to 4.109.}$$

Table A21: GHG Emissions from FT Production
(Per bbl of FT Liquid Product)

	Case 1	Case 2	Case 3	Case 4	Case 5*	Case 6*	Case 7	Case 8*
Feedstock	IL #6	IL #6	IL #6	Wyo. Coal	Biomass	Pipeline Gas	Assoc. Gas	Assoc. Gas
Upgrading	Maximum Distillate	Increased Gasoline	Maximum Gaso. & Chem.	Maximum Distillate	Fuels & Power	Maximum Distillate	Minimum Upgrading	Min. Upgrading & Power
CO_2 – vented gas (g)	443800	441652	400060	440972	0	64289	94294	0
CO_2 – combustion flue gas (g)	47685	44538	65931	92081	706987	54565	115726	92978
CO_2 – incineration flue gas (g)	17803	17739	16037	5493	0	0	0	0
CO_2 – fugitive emissions (g)	5105	5081	4601	5126	0	643	943	0
CO_2 – ancillary sources (g)	19917	17675	20530	31531	0	191	0	0
CH_4 – combustion flue gas (g CO_2-eq)	15	12	14	15	225	22	28	43
CH_4 – fugitive & flaring (g CO_2-eq)	145	145	145	145	47	141	73	57
CH_4 – ancillary sources (g CO_2-eq)	1070	917	1193	1673	0	14	0	0
N_2O – combustion flue gas (g CO_2-eq)	331	266	328	334	5115	497	626	981
N_2O – ancillary (g CO_2-eq)	337	325	327	551	0	0	0	0
Total (g CO_2-eq)	536209	528350	509166	577921	712374	120361	211690	94060

* Values reported only include allocation to fuel production

SECTION 4

RESOURCE
EXTRACTION

4. RESOURCE EXTRACTION

A. Utility consumption for coal production

STEP 1: Data Collection

**Table A22: Surface Coal Mining Utility and
Chemical Requirements
[16]**

		Units		Units
Electricity	14,300	MWh/year/MM tonne	44,311	Btu/ton
Fuel & Oil	269	m³/year/MM tonne	0.0645	Gal/ton
Ammonium Nitrate	2070	Mg/year/MM tonne	4.14	Lb/ton

Table A23: Underground Coal Mining Utility and
Chemical Requirements
[16]

		Units		Units
Electricity	12,755	MWh/year/MM tonne	39,523	Btu/ton
Raw Water	84,482	m³/year/MM tonne	20.3	Gal/ton
Limestone	16,263	Mg/year/MM tonne	32.5	Lb/ton

Table A24: Coal Cleaning Utility and
Landfilling Requirements (Base Case)
[16]

		Units
Electricity	0.79	MJ/Mg of MAF raw coal
Raw Water	0.17	m³/Mg of raw coal
Refuse	0.35	Dry Mg/Mg of MAF raw coal

STEP 2: Calculate the resource consumption for coal production using Tables A22, A23 and A24.

*Note: MF = Moisture Free
MAF = Moisture & Ash Free

*Illinois #6 underground coal contains 11.5% ash and Wyoming coal contains 8.7% ash. See
Ultimate analysis Table A36 at end of Appendix A.

Refuse:
Refuse includes ash forming material, rocks and very fine coal that are removed during coal
cleaning. Ultimate analysis coal data is moisture free, therefore subtract % ash from 1.0 to obtain
moisture free & ash free coal basis. Equation 4.1 is based on percentages, therefore any units can
be used such as lb refuse/lb MF coal produced or ton refuse/ton MF coal produced.

$$refuse\ (Ton\ /\ TonMFcoalp\ roduced\ ) = \left(\frac{0.35\,Mgrefuse}{MgMAFrawco\ al} \right) \times \left(\frac{MAFrawcoal}{MFrawcoal} \right) \qquad \text{(Eq 4.1)}$$

$$refuse\ (Ton\ /\ TonMFcoalp\ roduced\ ) = \left(\frac{0.35\,Mgrefuse}{MgMAFrawco\ al} \right) \times \left(\frac{(1-0.115)\,MAFrawcoal}{1.0\,MFrawcoal} \right) = 0.3098$$

Water Make-Up:

Water Make-up for the Illinois #6 case includes water for underground mining procedures <u>and</u> above ground coal cleaning processes.

Underground water: The underground water consumption is greater than 20.25 gallons/ton coal because refuse (ash and rocks) is included in the total coal mined until it reaches the cleaning/separation process. Therefore, the water consumption is based on the total bulk material removed underground.

$$H_2O(gal\ /\ Ton) = \left(\frac{20.25\,gal}{TonMFCoalp\ roduced} \right) \times \left(\frac{1-0.3098\,Tonrefuse}{TonMFCoalp\ roduced} \right) = 26.52 \qquad \text{(Eq 4.2)}$$

Coal cleaning water:

$$H_2O(gal\ /\ Ton) = \left(\frac{0.17\,m^3}{1MMtonMAFRawcoal} \right) \times 264\,gal\ /\ m^3 \times 907185g\ /\ Ton \times \left(\frac{1-0.115\,TonMAFRawcoal}{TonMFCoalproduced} \right) = 36.1 \qquad \text{(Eq 4.3)}$$

Water required per ton of coal produced:
H$_2$O Total (gal/ton) = 62.62

Limestone:

$$Limestone(ton\ /\ tonMFcoalproduced) = \left(\frac{32.54\,lbLimestone}{TonMFCoalproduced} \right) \times \left(\frac{1+0.3098}{TonMFCoalproduced} \right) \qquad \text{(Eq 4.4)}$$

Limestone Total (ton/toncoal) = 42.62

Electricity:

Electricity for the Illinois #6 case includes electricity for the underground coal extraction process <u>and</u> surface coal cleaning process (Jig washing).

Underground Electricity:

$$Electricity(kWh\ /\ ton) = \left(\frac{12755e6Wh}{TonneMFCoalproduced} \right) \times (1/1.102) \times \left(\frac{1+0.3098\,Tonrefuse}{TonMFCoalproduced} \right) \times \left(\frac{1}{1000} \right) = 15.35 \qquad \text{(Eq 4.5)}$$

Coal Cleaning Electricity:

$$Electricity(kWh\ /\ ton) = \left(\frac{0.79e6J}{1e6MAFrawcoal} \right) \times (1hr\ /\ 3600s) \times 907185g\ /\ Ton \times \left(\frac{1-0.115\,TonMAFrawcoal}{TonMFCoalproduced} \right) \times \left(\frac{1kW}{1000W} \right) = 0.176 \qquad \text{(Eq 4.6)}$$

Electricity Total (kWh/tonCoal) = 15.4

Table A25: Resource Consumption for Coal Production
(Per ton of MF Coal Produced)

	Illinois #6 Underground Mine	Illinois #6 Surface Mine	Wyoming Surface Mine
Electricity (kWh)	15.4	58.3	17.4
Distillate Fuel (gal)		0.084	0.089
Water Make-Up (gal)	62.62	46.06	44.65
Limestone (lb)	42.6		
Ammonium Nitrate (lb)		5.42	5.46
Refuse (ton)	-0.310	-0.310	-0.320

B. Coal Bed Methane

Coal bed methane is produced from the underground mining activities (extraction of coal to the surface) and underground post-mining activities (treatment of underground coal). The underground post-mining activities are not to be confused with surface strip mining. The post-mining activities include the handling, cleaning, etc. of the coal once it is brought to the surface. The EPA gives the post-mining methane emission factor (standard cubic feet of methane emitted per ton coal produced) directly, but the underground mining factor must be calculated from other EPA data.

Total Illinois Underground coal production (tons): 64,728,000   [18]
Total Illinois Underground methane (scf): 8,571e6   [18]
Illinois Underground <u>post mining</u> emission factor (scf/ton): 12.7*   [18]
*Post mining emission factor given directly by the EPA   [18]

Calculate the underground <u>mining</u> emission factor.

$$CH_4 Underground(scf/Tons) = \left(\frac{8,571e6 \, scfCH_4}{64,728e3 \, TonsCoalProduced} \right) = 132.4 \qquad \text{(Eq 4.7)}$$

$$CH_4 UndergroundTotal(scf/Ton) = CH_4 Underground + CH_4 UndergroundPost \qquad \text{(Eq 4.8)}$$

CH$_4$ Total (scf/ton) = 145 or (2779 g/ton)

Table A26: Coalbed Methane Emissions
(Per ton of MF Coal Produced)
[18]

	Illinois #6 Underground Mine	Illinois #6 Surface Mine	Wyoming Surface Mine
CH$_4$ (scf)	145	90	7.4
CH$_4$ (g)	2779	1725	142

C. Emissions inventory for coal production

Emissions sources included in the inventory are <u>coalbed methane release</u>, emissions from <u>electricity</u>, and emissions from <u>diesel fuel</u>. No ancillary diesel fuel is used for Design Case 1, Illinois #6 underground mining.

STEP 1: Calculate the emissions for each component

CO$_2$ emissions:

Source: Electricity (No diesel fuel is used in underground mining)

$$CO_2 Power(g/TonMFCoalproduced) = \left(\frac{15.35 kWhUsed}{TonMFCoalproduced}\right) \times \left(\frac{710.10 gCO_2}{kWhproduced}\right) = 10904 \qquad \text{(Eq 4.9)}$$

CH$_4$ emissions:

Source: Electricity and coalbed methane.

$$CH_4 Power(g/TonMFCoalproduced) = \left(\frac{15.35 kWhUsed}{TonMFCoalproduced}\right) \times \left(\frac{1.756 gCH_4}{kWhproduced}\right) = 26.9 \qquad \text{(Eq 4.10)}$$

CH$_4$ Coalbed Methane (g/tonMFCoalproduced) = 2779*
*Table A26

Methane Total (g/tonMFCoalproduced) = 2806

N$_2$O emissions:

Source: Electricity

$$N_2 OPower(g/TonMFCoalproduced) = \left(\frac{15.35 kWhUsed}{TonMFCoalproduced}\right) \times \left(\frac{0.0421 gN_2O}{kWhproduced}\right) = 0.646 \qquad \text{(Eq 4.11)}$$

SOx emissions:

Source: Electricity

$$SOxPower(g/TonMFCoalproduced) = \left(\frac{15.35 kWhUsed}{TonMFCoalproduced}\right) \times \left(\frac{6.92 gSOx}{kWhproduced}\right) = 106.2 \qquad \text{(Eq 4.12)}$$

NOx emissions:

Source: Electricity

$$NOxPower(g/TonMFCoalproduced) = \left(\frac{15.35 kWhUsed}{TonMFCoalproduced}\right) \times \left(\frac{1.81 gNOx}{kWhproduced}\right) = 27.8 \qquad \text{(Eq 4.13)}$$

CO emissions:

Source: Electricity

$$COPower(g/TonMFCoalproduced) = \left(\frac{15.35 kWhUsed}{TonMFCoalproduced}\right) \times \left(\frac{0.205 gCO}{kWhproduced}\right) = 3.15 \qquad \text{(Eq 4.14)}$$

VOC emissions:

$$VOCPower(g/TonMFCoalproduced) = \left(\frac{15.35kWhUsed}{TonMFCoalPproduced}\right) \times \left(\frac{1.81gVOC}{kWhproduced}\right) = 27.8 \qquad \text{(Eq 4.15)}$$

PM emissions:

$$PMPower(g/TonMFCoalproduced) = \left(\frac{15.35kWhUsed}{TonMFCoalproduced}\right) \times \left(\frac{1.91gPM}{kWhproduced}\right) = 29.3 \qquad \text{(Eq 4.16)}$$

Table A27: Emissions Inventory for Coal Production
(Per ton of MF Coal Produced)

	Illinois #6 Underground Mine	Illinois #6 Surface Mine	Wyoming Surface Mine
CO_2 (g)	10904	41425	12358
CH_4 (g)	2806	1826	172
N_2O (g)	0.65	2.5	0.73
SOx (g)	106.2	403	120.2
NOx (g)	27.8	105.2	31.6
CO (g)	3.2	12.1	3.7
VOC (g)	27.8	105.5	31.4
PM (g)	29.3	111.3	33.2

STEP 2: Convert the emissions inventory data (Table A27) for coal production into CO_2 equivalents using the global warming potential factors in Table A20 for methane and nitrous oxide.

$$CH_4(gCO_2 - eq/TonMFCoalproduced) = \left(\frac{2962gCH_4}{TonMFCoalproduced}\right) \times \left(\frac{21gCO2 - eq}{gCH_4}\right) = 62202 \qquad \text{(Eq 4.17)}$$

$$N_2O(gCO_2 - eq/TonMFCoalproduced) = \left(\frac{0.65gN_2O}{TonMFCoalproduced}\right) \times \left(\frac{310gCO2 - eq}{gN_2O}\right) = 200 \qquad \text{(Eq 4.18)}$$

Table A28: Greenhouse Gas Emissions from Coal Production
(Per ton of MF Coal Produced)

	Illinois #6 Underground Mine	Illinois #6 Surface Mine	Wyoming Surface Mine
CO_2 (g)	10904	12272	12358
CH_4 (g CO_2-eq)	58928	36850	3618
N_2O (g CO_2-eq)	200	225	227
Total	70030	49348	16203

A-47

SECTION 5

TRANSPORTATION
AND
DISTRIBUTION

5. TRANSPORTATION AND DISTRIBUTION

Design Case 1 of Scenario 1 coal is mined in southern Illinois and the FT plant is next to coal mine. The FT fuels produced are shipped by pipeline to the Chicago area (~200 miles) and distributed to a local re-fueling station by tank truck (~60 miles). The pipeline uses electricity and the tank truck uses distillate fuel. Emissions for both types of FT transportation (pipeline and tank truck) were calculated in the Section 2 "Ancillary Emissions".

A. Emissions Inventory for Transportation

STEP 1: Data collection

FT density (lb/gal):   6.163 lb/gal
Pipeline (miles):   200 miles
Tank truck (miles):   60 miles

Table A29: Energy Consumption for Different Modes of Transportation (Btu/ton-mile)
[20, pg. E-5]

Truck	Tanker	Barge	Train	Pipeline
1900	408	197	516	120

Table A30: Upstream and Combustion Emission Factors for Distillate Fuel, Residual Fuel and Electricity. (lb/MM Btu fuel consumed)
[Calculated in Ancillary Section]

	CO_2	CH_4	N_2O	SOx	NOx	CO	PM	VOC
Distillate Fuel	177	0.009	0.006	0.871	0.767	1.027	0.147	0.007
Residual Fuel	191	0.254	0.004	2.396	0.627	0.109	0.147	0.094
Electricity	149	0.368	0.008	1.45	0.55	0.176	0.40	0.004

STEP 2: Calculate the emissions per gallon of FT fuel transported (Pipeline to Chicago and then Chicago to distribution).

$$Emissions(g/galFTFuel) = (EnergyConsumption) \times (Distance) \times (Density) \times (Emissions) \times (Conv.Fact) \qquad \text{(Eq 5.1)}$$

CO_2 transportation example:

Truck:

$$TruckCO_2(g/galFT) = (1900) \times (60) \times (1/2000) \times (6.163) \times (177.4/1e6) \times (454) = 28.29 \qquad \text{(Eq 5.2)}$$

Pipeline:

$$PipelineCO_2\,(g/galFT) = (120)\times(200)\times(1/2000)\times(6.163)\times(148.9/1e6)\times(454) = 5.00 \qquad \text{(Eq 5.3)}$$

Total CO_2 (g/galFT) = 33.3

Methane Transportation Example:

Truck:
$$TruckCH_4\,(g/galFT) = (1900)\times(60)\times(6.163)\times(0.009471/1e6)\times(454) = 0.00151 \qquad \text{(Eq 5.4)}$$

Pipeline:
$$PipelineCH_4\,(g/galFT) = (120)\times(200)\times(1/2000)\times(6.163)\times(0.3684/1e6)\times(454) = 0.01237 \qquad \text{(Eq 5.5)}$$

Total Methane (g/gal FT) = 0.01388

Calculate the remaining component emissions using equation 5.1.

Table A31: Emissions Inventory for Transportation Scenario 1
(Per gal of FT Fuel Transported)

Transportation Mode		Truck	Tanker	Pipeline	Total
Southern Illinois to Chicago		DFO	RFO	Electricity	
	Miles	60	na	200	260
CO_2	(g)	28.29	na	5.00	33.3
CH_4	(g)	0.0015	na	0.0124	0.0139
N_2O (g)		0.0009	na	0.0003	0.0012
SOx (g)		0.1389	na	0.0487	0.1876
NOx	(g)	0.1223	na	0.0185	0.1408
CO	(g)	0.1638	na	0.0059	0.167
PM	(g)	0.0235	na	0.0134	0.0369
VOCs	(g)	0.0011	na	0.00013	0.0012

B. Greenhouse Gas Emissions from Transportation

Multiply the global warming potential factors (Table A20) by the transportation emissions inventory (Table A31). All scenarios presented in Table 32.

Table A32: Greenhouse Gas Emissions from Transportation
(Per gal of FT Fuel Transported)

	Truck	Tanker	Pipeline	Total
Scenario 1, 3 & 4 (g CO_2-eq/gal FT)	28.61	na	5.34	33.96
Scenario 2 (g CO_2-eq/gal FT)	28.61	na	26.74	55.35
Scenario 5 (g CO_2-eq/gal FT)	28.61	225.57	32.08	286.26
Scenario 6 (g CO_2-eq/gal FT)	28.61	465.80	21.39	516.80

SECTION 6

FULL FISCHER TROPSCH FUEL LIFE-CYCLE INVENTORY

6.   FULL FT-FUEL LIFE –CYCLE INVENTORY

Six baseline scenarios were identified for consideration in this study. They involve the evaluation of different options for the resource extraction, conversion, and transportation/distribution steps in the FT fuel chain. Detailed calculations of Scenario 1 are presented here.

Scenario 1: Production of FT fuels from bituminous Illinois No. 6 coal at a mine-mouth location in southern Illinois. The mine is an underground longwall mine. The design of the FT conversion plant is based on Design Case 1 described in Section 3 of the main report. Upgrading includes a full slate of refinery processes for upgrading FT naphtha. Hydrocracking is used to convert the FT wax into additional naphtha and distillate. The liquid fuel products are shipped by pipeline to a terminal in the Chicago area and distributed by tank truck to re-fueling stations in the immediate area.

A. Emissions Inventory for Full FT Fuel Chain

Individual inventories for the <u>FT conversion</u> (Section 3), <u>resource extraction</u> (Section 4), <u>and transportation/distribution</u> (Section 5) steps of the FT fuel chain are compiled here. They are the full inventories up through the point of sale of the FT fuel, and are based on the entire FT liquid-fuel product slate. That is, the individual products: LPG, gasoline/naphtha, and distillate fuel have not been broken out separately. Re-fueling and end-use combustion are <u>not</u> included. GHG emission allocation to diesel fuel only and combustion emissions are considered in the next case study. All values for Scenario 1 were calculated in the above sections. An example using carbon dioxide is shown below.

$$CO_2 (g/galFTFuel) = CO_2 Extraction + CO_2 Converstion + CO_2 Transportation \qquad \text{(Eq 6.1)}$$

STEP 1: Use data in Tables A16 and A27 to determine the airborne emissions from coal extraction per gallon of FT produced.

<u>CO_2 example:</u>

Data: Coal consumption (ton/bblFT) = 0.36745   [Table A16]
$\qquad\quad$ CO_2 (g/MF ton coal) = 10904   [Table A27]

$$CO_2 Extraction(gCO_2/galFT) = \left(\frac{10904 gCO_2}{toncoal}\right) \times \left(\frac{0.36745 toncoal}{bbl}\right) \times \left(\frac{bbl}{42 gal}\right) = 95.4 \qquad \text{(Eq 6.2)}$$

<u>Methane example:</u>

Data: Coal consumption (ton/bblFT) = 0.36745   [Table A14]
$\qquad\quad$ CH_4 (g/MF ton coal) = 2806   [Table A27]

$$CH_4 Extraction(gCO_2/galFT) = \left(\frac{2806 gCO_2}{toncoal}\right) \times \left(\frac{0.36745 toncoal}{bbl}\right) \times \left(\frac{bbl}{42 gal}\right) = 24.5 \qquad \text{(Eq 6.3)}$$

STEP 2: Calculate the Full FT Fuel Chain Emissions.

CO$_2$ example:

CO$_2$ FT Conversion (g/Bbl FT Product) = 534311   [Table A18]

CO$_2$ Transportation (g/gal FT Product) = 33.3   [Table A31]

Total CO$_2$ emissions for FT fuels at point of sale (use Eq 7.1):

$$CO_2(g/galFTFuel) = 95.4 + \left(\frac{534311}{42}\right) + 33.3 = 12857 \qquad \text{(Eq 6.4)}$$

Methane example:

CH$_4$ FT Conversion (g/Bbl FT Product) = 58.55   [Table A18]

CH$_4$ Transportation (g/gal FT Product) = 0.0139   [Table A31]

Total CH$_4$ emissions for FT fuels at point of sale:

$$CH_4(g/galFTFuel) = 24.5 + \left(\frac{58.55}{42}\right) + 0.0139 = 26.0 \qquad \text{(Eq 6.5)}$$

Calculate and tabulate the remaining emissions inventory for FT fuels at point of sale using data in Tables A16, A18, A27, A31 and equations 6.1 and 6.2.

Table A33: Emissions Inventory for FT Fuels at Point of Sale
(Per gal of FT Fuel Supplied)

		Scenario 1	Scenario 2	Scenario 3	Scenario 4	Scenario 5	Scenario 6
CO$_2$	(g)	12857	13865	-6564	4236	6385	6607
CH$_4$	(g)	26.0	3.76	0.45	14.9	6.07	6.36
N$_2$O	(g)	0.059	0.08	0.65	0.08	0.09	0.096
SOx	(g)	5.82	8.61	0.19	0.23	3.22	6.03
NOx	(g)	2.50	3.34	17.8	11.7	10.4	10.8
CO	(g)	0.57	0.68	5.33	2.98	2.46	2.49
VOC	(g)	1.73	2.47	2.66	16.5	13.2	13.2
PM	(g)	1.49	2.35	0.30	0.06	0.30	0.45

B. Case Study—Substitution of FT Diesel Fuel in SUVs

The results from the FT LCI were used to evaluate the application of FT diesel as a substitute for petroleum fuels in Sport Utility Vehicles (SUVs) and the greenhouse gas emissions that would result. FT diesel has been demonstrated to have emissions that are much lower than those from petroleum diesel for the same engine. There is however a penalty to fuel economy when using FT diesel due to its lower energy density per gallon to petroleum-derived diesel. FT diesel fuel economy in an SUV has been estimated to be about 24.4 mpg. The full life-cycle GHG emissions for FT diesel is presented here is based on Scenario 1, Illinois #6 coal.

Results include airborne emissions from <u>extraction/production</u>, <u>conversion/refining</u>, <u>transportation/distribution</u> and <u>end use combustion</u>. Results are given in g CO_2-equivalent per mile in SUV.

STEP 1: Determine the FT diesel allocation by using data in Table A14. Divide FT diesel produced by total FT liquid produced.

$$DieselAllocation = \left(\frac{24686}{24686 + 1922 + 23943} \right) = 0.49 \qquad \text{(Eq 6.6)}$$

STEP 2: Calculate airborne emissions per SUV mile from coal <u>extraction</u>.

Data: Coal consumption (ton/bblFT) = 0.36745   [Table A16]
        GHG emissions from coal production (gCO$_2$-eq/ton) = 70,032   [Table A28]

$$Extraction\,(gCO_2eq\,/\,SUVmile) = \left(\frac{.51}{.49} \right) \times \left(\frac{70032\,gCO_2eq}{toncoal} \right) \times \left(\frac{0.36745\,toncoal}{bblFT} \right) \times \left(\frac{bblFT}{42\,gal} \right) \times \left(\frac{galFT}{24.4\,SUVmiles} \right) = 26.1 \qquad \text{(Eq 6.7)}$$

STEP 3: Calculate airborne emissions per SUV mile for <u>conversion/refining</u>.

Data: GHG emissions from FT production (gCO$_2$-eq/bblFT) = 536,209   [Table A21]

$$Conversion\,(gCO_2eq\,/\,SUVmile) = \left(\frac{.51}{.49} \right) \times \left(\frac{536209\,gCO_2eq}{bblFT} \right) \times \left(\frac{bbl}{42\,gal} \right) \times \left(\frac{gal}{24.4\,mile} \right) = 543 \qquad \text{(Eq 6.8)}$$

STEP 4: Calculate airborne emissions per SUV mile for <u>transportation/distribution</u>.

Data: GHG emissions from Trans/Dist (gCO$_2$-eq/galFT) = 33.96   [Table A32]

$$Transportation\,(gCO_2eq\,/\,SUVmile) = \left(\frac{.51}{.49} \right) \times \left(\frac{33.96\,gCO_2eq}{galFT} \right) \times \left(\frac{gal}{24.4\,mile} \right) = 1.45 \qquad \text{(Eq 6.9)}$$

STEP 5: Calculate airborne emissions for end use <u>combustion</u> of FT diesel fuel.

Data: Combustion (gCO$_2$/gal FT fuel) = 9011.05   [Table A36]

$$Combustion\,(gCO_2eq\,/\,SUVmile) = \left(\frac{9011.05\,gCO_2}{galFT} \right) \times \left(\frac{gal}{24.4\,mile} \right) = 368 \qquad \text{(Eq 6.10)}$$

STEP 6: Aggregate the Total Fuel Chain GHG Emissions.

$$Total(gCO_2eq\,/\,SUVmile) = 26.0 + 543 + 1.45 + 368 = 939 \qquad \text{(Eq 6.11)}$$

Table A34: Full Life-Cycle GHG Emissions for FT Diesel
(g CO_2-eq/SUV mile)

Scenario/ FT Plant Feedstock	Extraction/ Production	Conversion/ Refining	Transportation/ Distribution	End Use Combustion	Total
1) IL #6 Coal	26	543	1.4	368	939
2) Wyoming Coal	7	585	2.3	368	962
3) Plantation Biomass	-969	703	1.4	368	104
4) Pipeline Natural Gas	71	121	1.4	368	562
5) Venezuelan Assoc. Gas	51	213	11	368	643
6) ANS Associated Gas	51	213	21	368	652

C. Sensitivity Cases for Substitution of FT Diesel Fuel in SUVs

To help identify possible GHG reduction strategies for FT fuels production, a number of sensitivity cases were considered for the scenarios described above. These included the following:

- Advanced diesel engines
- Coalbed methane capture
- Sequestration of vented CO_2 from conversion process
- Sequestration of CO_2 from conversion process and combustion
- Co-production of fuels and power
- Co-processing of coal and biomass
- Co-processing of coal and coalbed methane

Re-calculate the Full Life-Cycle GHG emissions based on SUV miles as shown in the previous section but with taking into account the reduction scenarios.

1a). Illinois #6 coal baseline
Total fuel chain emissions from Table A34 above is 939 g CO_2-eq/mile in SUV.

1b). Sequestration of FT process CO_2
This involves re-calculating the airborne emissions for the FT conversion process, minus the vented CO_2 emissions.

Data: Total FT process CO_2 (gCO_2-eq/bblFT) = 536209   [Table A21]
       Vented CO_2 (gCO_2-eq/bblFT) = 443800   [Table A21]
      *The remaining extraction, transportation and combustion emissions remain unchanged.

Re-calculated FT conversion emissions:

$$Conversion\ (gCO_2eq\ /\ SUVmile) = \left(\frac{.51}{.49}\right) \times \left(\frac{536209 - 443800}{42 \times 24.4}\right) = 93.9 \qquad \text{(Eq 6.12)}$$

Re-calculated existing diesel engine fuel chain emissions:
$$Total\ (gCO_2eq\ /\ SUVmile) = 26 + 93.9 + 1.4 + 368 = 490 \quad \text{(Compared to 939!)} \qquad \text{(Eq 6.13)}$$

Reduction amount = 449 gCO_2-eq/SUVmile or 48%.

*Assume <u>advanced</u> diesel engine has 13% lower emissions than <u>existing</u> diesel engine.

$$AdvancedDiesel(gCO_2eq \, / \, SUVmile) = 490 \times (1-.13) = 426 \qquad \text{(Eq 6.14)}$$

1c). <u>Sequestration of Vented and Combusted GHG Emissions</u>

This involves re-calculating the airborne emissions for the FT <u>conversion</u> process, minus the emissions from: vented CO_2, CO_2 combustion flue gas, CH_4 combustion flue gas, and N_2O combustion flue gas.

Data:
Total FT process CO_2 (gCO$_2$-eq/bblFT) = 536209	[Table A21]
Vented CO_2 (gCO$_2$-eq/bblFT) = 443800	[Table A21]
CO_2 combustion (gCO$_2$-eq/bblFT) = 47685	[Table A21]
CO_2 incineration (gCO$_2$-eq/bblFT) = 17803	[Table A21]
CH_4 combustion (gCO$_2$-eq/bblFT) = 15	[Table A21]
N_2O combustion (gCO$_2$-eq/bblFT) = 331	[Table A21]

*The remaining extraction, transportation and end-use combustion emissions remain unchanged.

Re-calculated FT conversion emissions:

$$Conversion\,(gCO_2eq \, / \, SUVmile) = \left(\frac{.51}{.49}\right) \times \left(\frac{536809 - 443800 - 47685 - 17803 - 15 - 331}{42 \times 24.4}\right) = 27 \qquad \text{(Eq 6.15)}$$

Re-calculated existing diesel engine fuel chain emissions:

$$Total(gCO_2eq \, / \, SUVmile) = 26 + 27 + 1.4 + 368 = 423 \qquad \text{(Eq 6.16)}$$

Reduction amount = 516 gCO$_2$-eq/SUVmile or 55%.

*Assume advanced diesel engine has 13% lower emissions than existing diesel engine.

$$AdvancedDiesel(gCO_2eq \, / \, SUVmile) = 423 \times (1-.13) = 368 \qquad \text{(Eq 6.17)}$$

1d). <u>Co-production of fuels and power</u>

Plant efficiency improvements due to this "once-through" conversion approach results in a 56% reduction in emissions from FT production (conversion). The remaining extraction, transportation and combustion emissions remain unchanged from the baseline.

Re-calculated FT conversion emissions:

$$Conversion(gCO_2eq \, / \, SUVmile) = 543 \times (1-.56) = 239 \qquad \text{(Eq 6.18)}$$

Re-calculated existing diesel engine fuel chain emissions:

$$Total(gCO_2eq \, / \, SUVmile) = 26 + 239 + 1.4 + 368 = 635 \qquad \text{(Eq 6.19)}$$

Reduction amount = 304 gCO$_2$-eq/SUVmile or 32%.

*Assume advanced diesel engine has 13% lower emissions than existing diesel engine.

$$AdvancedDiesel(gCO_2eq \, / \, SUVmile) = 635 \times (1-.13) = 552 \qquad \text{(Eq 6.20)}$$

1e). Co-processing of biomass

Co-processing of other feedstocks with coal may also be a viable approach to reducing GHG emissions. Here are results of co-feeding 20% of the feedstock from biomass (based on heating value).

Data:
    Coal LHV = 11945 Btu/lb or 23.89 MM Btu/ton   [7]

    Biomass LHV = 1124 Btu/lb or 15.44 MM Btu/ton   [7]

    Basis (MM Btu) = 100 (80 MM to coal, 20 MM to bio)

    Coal (ton/bbl FT liquid product) = 0.3675   [Table A16]

    Biomass (ton/bbl FT liquid product) = 0.621   [Table A16]

With the given data, it was determined that 3.3486 tons of coal and 1.2953 tons of biomass are required for each 100 MM Btu feedstock to the gasifier.

$$CoalConv \ (galFT \ / \ 80 \ MMBtu) = \left(\frac{3.3486 \ ton \ coal}{80 \ MMBtu} \right) \times \left(\frac{bbl \ FT}{0.36745 \ ton \ coal} \right) \times \left(\frac{42 \ gal}{bbl \ FT} \right) = 382.7 \qquad \text{(Eq 6.21)}$$

$$BioConv \ (galFT \ / \ 20 \ MMBtu) = \left(\frac{1.2953 \ ton \ biomass}{20 \ MMBtu} \right) \times \left(\frac{bbl \ FT}{0.621 \ ton \ biomass} \right) \times \left(\frac{42 \ gal}{bbl \ FT} \right) = 87.6 \qquad \text{(Eq 6.22)}$$

$$\% \ fromBio = \left(\frac{87.6}{382.7 + 87.6} \right) \times 100 = 18.6 \qquad \text{(Eq 6.23)}$$

% From Coal = 81.4

Use the Scenario 1 (coal) baseline and Scenario 3 (biomass) data in Table A34 and the allocated percentages for biomass and coal to re-calculate the full life-cycle GHG emissions for the entire fuel chain; extraction, conversion, transportation and end use combustion.

Re-calculated biomass and coal extraction emissions:

$$Extraction \ (gCO_2 - eq \ / \ SUVmile) = (26 \times .814) + (-969 \times 0.186) = -159 \qquad \text{(Eq 6.24)}$$

Re-calculated biomass and coal conversion emissions:

$$Conversion \ (gCO_2 - eq \ / \ SUVmile) = (543 \times .814) + (703 \times 0.186) = 572 \qquad \text{(Eq 6.25)}$$

Re-calculated biomass and coal transportation emissions:

$$Transportation \ (gCO_2 - eq \ / \ SUVmile) = (1.388 \times .814) + (1.456 \times 0.186) = 1.4 \qquad \text{(Eq 6.26)}$$

*Assume no change in end-use combustion.

Re-calculated existing diesel engine fuel chain emissions:

$$Total \ (gCO_2 eq \ / \ SUVmile) = -159 + 572 + 1.4 + 368 = 783 \qquad \text{(Eq 6.27)}$$

Reduction amount = 155 gCO$_2$-eq/SUVmile or 17%.

*Assume advanced diesel engine has 13% lower emissions than existing diesel engine.

$$AdvancedDiesel \ (gCO_2 eq \ / \ SUVmile) = 783 \times (1 - .13) = 682 \qquad \text{(Eq 6.28)}$$

1f). Coalbed methane capture

This involves re-calculating the airborne emissions for the coal extraction process, minus the coalbed methane. The remaining conversion, transportation and combustion values remain unchanged from the baseline (1a).

Data: Coalbed methane (gCH$_4$/toncoal) = 2779   [Table A26]
Coal consumption (ton/bblFT) = 0.36745   [Table A16]

Re-calculate CO_2 equivalent emissions from coalbed methane:

$$CH_4(gCO_2eq/bblFT) = \left(\frac{2779gCH_4}{toncoal}\right) \times \left(\frac{21gCO_2-eq}{gCH_4}\right) \times \left(\frac{0.36745toncoal}{bblFT}\right) = 21,444 \qquad \text{(Eq 6.29)}$$

Re-calculate total underground mining CO_2 equivalent emissions per bbl FT:

$$Total(gCO_2eq/bblFT) = \left(\frac{70032gCO_2eq}{toncoal}\right) \times \left(\frac{0.36745toncoal}{bblFT}\right) = 25,733 \qquad \text{(Eq 6.30)}$$

Re-calculate the extraction emissions (minus the coalbed methane):

$$Extraction(gCO_2eq/bblFT) = \left(\frac{.51}{.49}\right) \times \left(\frac{25,733-21,444gCO_2eq}{bblFT}\right) \times \left(\frac{bbl}{42gal}\right) \times \left(\frac{gal}{24.4mile}\right) = 4.3 \qquad \text{(Eq 6.31)}$$

Re-calculate existing diesel engine fuel chain emissions:

$$Total\,(gCO_2eq/SUVmile) = 4.3 + 543 + 1.4 + 368 = 917 \qquad \text{(Eq 6.32)}$$

Reduction amount = 22 gCO$_2$-eq/SUVmile or 2.3%.

*Assume advanced diesel engine has 13% lower emissions than existing diesel engine.

$$AdvancedDiesel(gCO_2eq/SUVmile) = 917 \times (1-0.13) = 798 \qquad \text{(Eq 6.33)}$$

1g). Co-processing of coalbed methane

Co-processing of coalbed methane involves re-calculating the airborne emissions for the full fuel chain by producing 50 percent of the FT product from methane and 50 percent of the FT product form coal. Extraction and conversion are different than the baseline case but transportation and combustion are assumed to be the same as the baseline since the FT products from co-processing are assumed to be similar to the FT products from the baseline scenario.

Scenario 1f emissions are used for the coal feedstock portion (50 percent) and Scenario 4a (modified pipeline gas) is used for the coalbed methane feedstock portion. A straight 50 percent of Scenario 1f emissions is allocated to the coal portion here for extraction and conversion. Fifty percent of Scenario 4a (pipeline gas) emissions are allocated to the coalbed methane portion here for conversion, but not for extraction. A pipeline gas transmission credit is subtracted from the extraction step since the FT plant is near the coal mine, and therefore, no gas transportation is required. This transmission credit is estimated to be 20gCO$_2$eq/SUVmile. A second credit from gas processing subtracted from the extraction step of the pipeline gas since the coalbed methane is not processed. The gas processing credit is estimated to be approximately 49 gCO$_2$eq/SUVmile. Note that these are only ESTIMATES!

Re-calculate the extraction emissions:

$$Extraction(gCO_2eq/SUVmile) = (0.5 \times 4.3) + (0.5 \times (71-20-49)) = 3.2 \qquad \text{(Eq 6.34)}$$

Re-calculated biomass and coal conversion emissions:

$$Conversion\,(gCO_2eq\,/\,SUVmile) = (0.5 \times 543) + (0.5 \times 121) = 332 \qquad \text{(Eq 6.35)}$$

Re-calculate existing diesel engine fuel chain emissions:
$$Total\,(gCO_2eq\,/\,SUVmile) = 3.2 + 332 + 1.4 + 368 = 705 \qquad \text{(Eq 6.36)}$$

Reduction amount = 234 gCO$_2$-eq/SUVmile or 25%.

*Assume advanced diesel engine has 13% lower emissions than existing diesel engine.
$$AdvancedDiesel\,(gCO_2eq\,/\,SUVmile) = 705 \times (1 - 0.13) = 613 \qquad \text{(Eq 6.37)}$$

Table A35: Life-Cycle Sensitivity Analysis for FT Diesel
(g CO$_2$-eq/SUV mile)

Scenario/ FT Feedstock Source	GHG Emissions Reduction		Total Fuel Chain	
			existing diesel engine	advanced diesel engine
1a) IL #6 coal - base case	-	-	939	816
1b) with seq. of process CO$_2$	449	48%	490	426
1c) with seq. of process & comb. CO$_2$	516	55%	423	368
1d) with co-prod. of fuels & power	304	32%	635	552
1e) with co-proc. of biomass	155	17%	783	682
1f) with coalbed CH$_4$ capture	22	2.3%	917	798
1g) with co-proc. of coalbed CH$_4$	234	25%	705	613
4a) Pipeline natural gas - base case	-	-	562	489
4b) with seq. of process CO$_2$	65	12%	497	432
4c) with seq. of process & comb. CO$_2$	120	22%	442	384
5a) Venezuelan assoc. gas - base case	-	-	643	559
5b) with flaring credit	578	90%	65	57
5c) with venting credit	3234	503%	-2592	-2255
6a) ANS associated gas _ base case	-	-	652	567
6b) with seq. of process CO$_2$	94	14%	558	485
6c) with seq. of process & comb. CO$_2$	211	32%	441	383
6d) with co-prod. of fuels & power	119	18%	534	464

Table A36: Ultimate Analysis

Table A36: Ultimate Analysis	HHV (MF) M Btu/lb	LHV (MF) M Btu/lb	% Moisture	% Ash (MF)	% C (MF)	% H (MF)	% N (MF)	% S (MF)	% Cl (MF)	% O (MF)	% Total (MF)	g CO₂/gal	g CO₂/ton	g CO₂/Mscf	g CO₂-eq/Mscf
IL#6 Coal (Burning Star Mine)	12.246	11.945	8.60	11.49	71.01	4.80	1.40	3.19	0.10	8.01	O by diff.				
IL#6 Slag (Shell Gasifier)				95.04	3.36	0	0	1.44	0.16		100.00				
Wyo Coal (Powder River Basin)	11.645	11.198		8.71	67.84	4.71	0.94	0.58	0.01	17.21	O by diff.				
Wyo Slag (Shell Gasifier)				95.04	3.36	0	0	1.44	0.16		100.00				
SRWC (Maple Wood Chips)	8.083	7.724	37.9	0.50	49.54	6.11	0.10	0.02	0.00	43.73	100.00		1646900.67		
Biomass Slag (BCL Gasifier)				3.25	89.20	7.48	0.00	0.07	0.00	0.00	100.00				
Pipeline Natural Gas	23.077	20.823	0	0	73.75	23.97	0.95	nil	0	1.33	100.01				
Associated Gas (xx% CO2)	17.021	15.367	0	0	61.96	17.59	0.00	nil	0.00	20.45	100.00			55983.549	313521.43
Fuel Gas (Case 1)	5.18	4.74			36.54	6.02	17.53	0.00	0.00	39.92					
Fuel Gas (Case 4)	7.45	6.90									0.00				
S-Plant Flue Gas (Case 1,2,3)					24.93	4.25	0.86	0.05095	0.00	69.91	100.00				
S-Plant Flue Gas (Case 4)					23.80	2.67	9.98	0.03	0.00	63.52	100.00				
Hydrogen (H2)	61.0	51.6				100.00									
Nitrogen (N2)	0.0	0.0					100.00								
Carbon Monoxide (CO)	4.3	4.3			42.88					57.12	100.00				
Carbon Dioxide (CO2)	0.0	0.0			27.29					72.71	100.00				
Carbonyl Sulfide (COS)	4.0	4.0			19.99			53.37		26.64	100.00				
Water (H2O)	0.0	0.0				11.19				88.81	100.01				
Hydrogen Sulfide (H2S)	7.1	6.5				5.92		94.07			99.99				
Ammonia (NH3)	9.7	8.0				17.76	82.27				100.02				
Hydrogen Cyanide (HCN)					44.43	3.73	51.83				99.99				
Methanol (CH3OH)	9.8	8.6			37.48	12.58				49.94	100.01				
MTBE (C5H12O)	16.3	15.0			68.12	13.72				18.15	100.00				
TAME (C6H14O)	17.0	15.7			70.52	13.81				15.66	99.99				
Methane (CH4)	23.9	21.5			74.88	25.14					100.01				
Ethylene (C2H4)	21.3	20.3			85.63	14.37					100.01				
Ethane (C2H6)	22.3	20.4			79.88	20.11					99.99				
Propylene (C3H6)	21.0	19.7			85.62	14.37					100.00				
LPG (Propane - C3H8)	21.7	19.931			81.72	18.29					100.01				
Butanes (C4H10)	21.3	19.634			82.66	17.34					100.00				
Pentanes (C5H12)	20.9	19.3			83.23	16.77					99.99				
Hexanes (C6H14)	20.8	19.2			83.63	16.38					100.00				
95 RONC Reformate	17.6	16.8			88.11	11.60					99.71				

LCA Inventory

Table A36: Ultimate Analysis	HHV (MF)	LHV (MF)	% Moisture	% Ash (MF)	% C (MF)	% H (MF)	% N (MF)	% S (MF)	% Cl (MF)	% O (MF)	% Total (MF)	g CO_2/gal	g CO_2/ton	g CO_2/Mscf	g CO_2-eq/Mscf
	M Btu/lb	M Btu/lb													
C5/C6 Isomerate (81 R+M/2)	20.1	18.5			83.44	16.49					99.93				
C3/C4/C5 Alkylate (92 R+M/2)	20.0	18.4			84.00	18.09					102.09				
ZSM-Gasoline	18.6	17.3			85.88	13.58					99.46				
Case 1 Gasoline	19.0	17.740			85.63	14.99					100.62	8551.98			
Case 2 Gasoline	19.4	17.962			85.05	15.35					100.41	8408.87			
Case 3 Gasoline	18.3	16.983			78.73	15.27				6.75	100.75	7825.33			
Case 4 Gasoline	19.0	17.741			85.63	14.99					100.62	8550.66			
Case 5 Gasoline	18.3	17.274			86.81	12.96					99.77	8813.61			
Case 6 Gasoline	18.8	17.610			85.95	14.39					100.34	8602.60			
FT-Derived Naphthas (C7-350°F)	20.7	19.100			84.60	15.40						8058.68			
FT-Derived Distillates (350°F+)	20.5	18.900			84.60	15.40						9011.05			
Case 3 Distillate	20.1	18.580			84.86	15.04						8956.28			

Appendix B:

Greenhouse-Gas Emissions Inventory Tables
Metric Units

Executive Summary Table
Full Life-Cycle GHG Emissions for FT & Petroleum Diesel
(g CO_2-eq/kilometer in SUV)

Resource	Extraction/ Production	Conversion/ Refining	Transport./ Distribution	End Use Combustion	Total Fuel Chain
IL #6 Coal - base case	16	337	1	229	583
- in advanced diesel*	37	293	1	199	507
Wyoming Coal	4	364	2	229	598
Plantation Biomass	-602	437	1	229	65
Pipeline Natural Gas	44	75	1	229	349
Venezuelan Assoc. Gas	32	132	7	229	400
- with flaring credit*	-327	132	7	229	40
ANS Associated Gas	32	132	13	229	405
Wyoming Sweet Crude Oil	14	46	5	226	291
Arab Light Crude Oil	22	50	16	228	316
ANS Crude Oil	17	63	9	235	324
Venezuelan Syncrude	20	89	6	242	357

*1.6093 kilometers = 1 mile

Table 1: Global Warming Potentials for Selected Gases
(kg of CO_2 per kg of Gas)

Gas	Lifetime (years)	Direct Effect over Time Horizons of:		
		20 Years	100 Years	500 Years
Carbon Dioxide (CO_2)	Variable	1	1	1
Methane (CH_4)	12 ± 3	56	21	7
Nitrous Oxide (N_2O)	120	280	310	170

Table 2: Indirect Liquefaction Baseline Design Data

Design	Option 1	Option 2	Option 3	Option 4	Option 5	Option 6	Option 7	Option 8
Feedstock	IL #6	IL #6	IL #6	Wyo. Coal	Biomass	Pipeline Gas	Assoc. Gas	Assoc. Gas
Upgrading	Maximum Distillate	Increased Gasoline	Maximum Gaso. & Chem.	Maximum Distillate	Fuels & Power	Maximum Distillate	Minimum Upgrading	Min. Upgrading & Power
Raw Materials (tonne/day)								
Coal, Biomass, NG	16851	16851	16851	17953	2000	8119	12502	12502
Catalysts & Chemicals	310	348	na	357	na	2.65	na	na
Products (liters/day)								
Methanol			-366153					
Propylene			804489					
LPG	305579	417031	250091	303194	0	270919	0	0
Butanes	-494459	158672	-827383	-493028	0	-54057	0	0
Gasoline/Naphtha	3806698	4969232	6315401	3776966	60734	2707123	2448446	1923779
Distillates	3924827	2521263	1552378	3889849	123217	4167287	5373862	4245033
Products (tonne/day)								
Methanol			-291					
Propylene			417					
LPG	155	211	127	153	0	137	0	0
Butanes	-287	92	-482	-287	0	-32	0	0
Gasoline/Naphtha	2741	3542	4525	2719	44	1953	1681	1320
Distillates	3033	1961	1181	3006	95	3213	4126	3253
By-Products (tonne/day)								
Slag	2036	2036	2036	1585	209			
Sulfur	508	459	459	98				
CO_2 Removal	25804	25777	25822	25696		2967	4639	
CO_2 Carrier Gas	-3370	-3370	-3370	-3591				
S-Plant Flue Gas	985	985	985	316				
Utilities Consumed								
Electric Power (MW)	54.3	53	58	88	-86	-25	0	-372
Raw Water (m³/day)	52996	52996	60567	37854	7571	79494	22713	15142

1 ton = 0.9072 tonne; 1 bbl = 158.99 liters; 1 m³ = 264 gallons

Table 3: Resource Consumption and Yields for FT Production
(Per m^3 of FT Liquid Product)

Design	Option 1	Option 2	Option 3	Option 4	Option 5	Option 6	Option 7	Option 8
Feedstock	IL #6	IL #6	IL #6	Wyo. Coal	Biomass	Pipeline Gas	Assoc. Gas	Assoc. Gas
Upgrading	Maximum Distillate	Increased Gasoline	Maximum Gaso. &Chem.	Maximum Distillate	Fuels & Power	Maximum Distillate	Minimum Upgrading	Min. Upgrading & Power
Resources								
Coal or Biomass (MF tonne)	2.10	2.09	1.89	2.25	3.54 [0.0041]			
Butanes (liter)	62		93	62		8		
Methanol (liter)		41	41					
Catalysts & Chemicals (kg)	358.7	440.5	na	448.2	na	3.7	na	na
Water Make-Up (m^3)	6.81	6.79	6.64	4.67	12.88 [0.0150]	10.83 [0.0220]	2.71	2.17 [0.0025]
Electric Power (kJ)	584292	563449	563094	953660	-40324528	-298868		-5207547
Volume Yield (liter)								
C3/C4 LPG	38	71	118	38		38		
Gasoline/Naphtha	474	616	708	474	330	379	313	312
Distillates	488	313	174	488	670	583	687	688
Mass Yield (tonne)								
C3/C4 LPG	0.0171	0.0403	0.6270	0.0170		0.0167		
Gasoline/Naphtha	0.3421	0.4396	0.5075	0.3421	0.2396	0.2736	0.2170	0.2170
Distillates	0.3767	0.2453	0.1314	0.3767	0.5195	0.4509	0.5252	0.5252
Slag (MF)	0.2509	0.2509	0.2283	0.20	0.3711			
Sulfur	0.6270	0.6270	0.5704	0.0113				
Energy Yield (MJ)								
C3/C4 LPG	893	1736	2799	887		887		
Gasoline/Naphtha	14069	18340	20031	14075	9710	11195	9547	9509
Distillates	16591	10503	5723	16579	22742	19767	23195	23189
Power					67207	849		8686
Allocation to Fuels					32.6%	97.4%		79.0%
Thermal Efficiency (LHV)	50.4%	52.0%	47.4%	49.3%	51.0%	59.1%	57.3%	57.1%
Carbon Efficiency	40.1%	41.1%	37.7%	39.1%	37.2%	57.0%	39.3%	39.2%

1 ton = 0.9072 tonne; 1 bbl = 158.99 liters; 1 bbl = .15899 m^3; 1 lb = 0.4536 kg; 1 Btu = 1055.1 joules; MJ = megajoule

Table 4: Emissions Inventory for FT Production
(Per liter of FT Liquid Product)

Design	Option 1	Option 2	Option 3	Option 4	Option 5*	Option 6*	Option 7	Option 8*
Feedstock	IL #6	IL #6	IL #6	Wyo. Coal	Biomass	Pipeline Gas	Assoc. Gas	Assoc. Gas
Upgrading	Maximum Distillate	Increased Gasoline	Maximum Gaso. & Chem.	Maximum Distillate	Fuels & Power	Maximum Distillate	Minimum Upgrading	Min. Upgrading & Power
CO_2 (mg)	3360658	3312689	3189880	3617859	4446737	752797	1326899	584803
CH_4 (mg)	368	322	405	549	82	53	30	30
N_2O (mg)	14	12	13	18	104	10	13	20
SO_x (mg)	1243	1200	1219	1875	0	0.4	0	0
NO_x (mg)	560	453	618	747	3295	327	404	632
CO (mg)	99	74	113	120	800	79	98	154
VOC (mg)	386	291	479	573	141	24	17	27
PM (mg)	317	303	312	513	71	7	9	14

Table 5: Emissions Inventory for Power Exported from FT Plants
(Per MJ* of Electric Power)

Design	Option 5	Option 6	Option 8
Feedstock	Biomass	Pipeline Gas	Assoc. Gas
Upgrading	Fuels & Power	Maximum Distillate	Min. Upgrading & Power
CO_2 (mg)	228333	67500	29722
CH_4 (mg)	4.2	4.7	1.7
N_2O (mg)	5.3	0.833	1.1
SOx (mg)	0	0	0
NOx (mg)	170	29.2	32.2
CO (mg)	41.1	7.2	7.8
VOC (mg)	7.2	2.2	1.4
PM (mg)	3.6	0.56	0.56

*MJ = megajoule = 1e6 joules

Table 6: GHG Emissions from FT Production
(Per liter of FT Liquid Product)

Design	Option 1	Option 2	Option 3	Option 4	Option 5*	Option 6*	Option 7	Option 8*
Feedstock	IL #6	IL #6	IL #6	Wyo. Coal	Biomass	Pipeline Gas	Assoc. Gas	Assoc. Gas
Upgrading	Maximum Distillate	Increased Gasoline	Maximum Gaso. & Chem.	Maximum Distillate	Fuels & Power	Maximum Distillate	Minimum Upgrading	Min. Upgrading & Power
CO_2 – vented gas (mg)	2791373	2777860	2516260	2773585	0	404356	593084	0
CO_2 – combustion flue gas (mg)	299928	280131	414687	579165	4446736	343198	727884	584803
CO_2 – incineration flue gas (mg)	111976	111573	100866	34549	0	0	0	0
CO_2 – fugitive emissions (mg)	32107	31957	28940	32241	0	4044	5931	0
CO_2 – ancillary sources (mg)	125271	111168	129127	198319	0	1198	0	0
CH_4 – combustion flue gas (mg CO_2-eq)	92	74	91	93	1417	138	173	272
CH_4 – fugitive & flaring (mg CO_2-eq)	912	912	912	912	297	888	456	360
CH_4 – ancillary sources (mg CO_2-eq)	6730	5769	7503	10522	0	90	0	0
N_2O – combustion flue gas (mg CO_2-eq)	2084	1676	2062	2101	32172	3124	3940	6172
N_2O – ancillary (mg CO_2-eq)	2122	2042	2055	3463	0	1	0	0
Total (mg CO_2-eq)	3372595	3323162	3202504	3634950	4480622	757037	1331468	591607

Table 8: Ultimate Analyses of Coal and Biomass

	Illinois #6 Coal	Wyoming Coal	Maplewood Chips
HHV (kJ/kg)	28494	27099	18795
LHV (kJ/kg)	27797	26052	17957
	Wt. %	Wt. %	Wt.%
Moisture	9.41	44.9	61.0
Ash	11.49	8.71	0.50
C	71.01	67.84	49.54
H	4.80	4.71	6.11
N	1.40	0.94	0.10
S	3.19	0.58	0.02
Cl	0.10	0.01	0.00
O (by diff.)	8.01	17.21	43.73

Table 9: Resource Consumption for Coal Production
(Per tonne of MF Coal Produced)

		Illinois #6 Underground Mine	Illinois #6 Surface Mine	Wyoming Surface Mine
Electricity	(kJ)	50120	56281	56674
Distillate Fuel	(liter)		0.290	0.292
Water Make-Up	(liter)	215	158	153
Limestone	(kg)	17.5		
Ammonium Nitrate	(kg)		2.23	2.25
Refuse	(tonnne)	-.310	-.310	-.310

*Positive value is consumed, negative is produced.

Table 10: Coalbed Methane Emissions
(Per tonne of MF Coal Produced)

		Illinois #6 Underground Mine	Illinois #6 Surface Mine	Wyoming Surface Mine
CH_4	(N liter)	3526	2188	180
CH_4	(mg)	2521232	1564471	128668

Table 11: Emissions Inventory for Coal Production
(Per tonne of MF Coal Produced)

		Illinois #6 Underground Mine	Illinois #6 Surface Mine	Wyoming Surface Mine
CO_2	(mg)	9892240	11133517	11211290
CH_4	(mg)	2545680	1591925	156314
N_2O	(mg)	586	659	663
SOx	(mg)	96341	108327	109083
NOx	(mg)	25060	28434	28633
CO	(mg)	2854	3331	3355
VOC	(mg)	25200	28318	28516
PM	(mg)	26591	29822	30091

Table 12: Greenhouse Gas Emissions from Coal Production
(Per tonne of MF Coal Produced)

		Illinois #6 Underground Mine	Illinois #6 Surface Mine	Wyoming Surface Mine
CO_2	(mg)	9892240	11133517	11211290
CH_4	(mg CO_2-eq)	53459274	33430431	3282593
N_2O	(mg CO_2-eq)	181698	204228	205655
Total	(mg CO_2-eq)	63533212	44768176	14699538

Table 13: Emissions Inventory for Biomass Production
(Per tonne of MF Biomass Produced)

		Feedstock Sequestering	Cultivation & Harvesting	Local Transportation	Total
CO_2	(g)	-1495281	47476	9219	-1438618
CH_4	(g)		7.55	0.35	7.9
N_2O	(g)		15.3	0.36	15.7
SOx	(g)		na	na	na
NOx	(g)		279	44.8	323
CO	(g)		112.5	18.1	130.6
VOC	(g)		117.3	13.3	130.6
PM	(g)		na	na	na

Table 14: Greenhouse Gas Emissions from Biomass Production
(Per tonne of MF Biomass Produced)

		Feedstock Sequestering	Cultivation & Harvesting	Local Transportation	Total
CO_2	(g CO_2)	-1495313	47477	9219	-1438618
CH_4	(g CO_2-eq)		159	7.4	166
N_2O	(g CO_2-eq)		4753	113	4865
Total	(g CO_2-eq)	-1495313	52388	9339	-1433586

Table 15: Composition of Associated & Pipeline Natural Gas

	Associated Gas	Pipeline Gas
HHV (kJ/N liter)	36.4	39.5
LHV (kJ/N Liter)	32.9	35.6
	Vol. %	Vol. %
Methane	76.2	94.7
Ethane	6.4	3.2
Propane	3.2	0.5
Isobutane	0.3	0.1
n-Butane	0.8	0.1
C_5+	0.1	0.1
CO_2	12.6	0.7
H_2S	-	-
N_2	0.4	0.6

Table 16: Emissions Inventory for Natural Gas Production
(Per Normal Liter of Natural Gas Produced)

		Associated Gas	Pipeline Gas
CO_2	(mg)	165	238
CH_4	(mg)	0.851	2.57
N_2O	(mg)	0.0056	0.0078
SOx	(mg)	na	0.0078
NOx	(mg)	1.26	1.81
CO	(mg)	0.3060	0.4403
VOC	(mg)	2.0	2.87
PM	(mg)	0	0

Table 17: Greenhouse Gas Emissions from Natural Gas Production
(Per Normal Liter of Natural Gas Produced)

		Associated Gas	Pipeline Gas
CO_2	(mg CO_2)	165	238
CH_4	(mg CO_2-eq)	18	54
N_2O	(mg CO_2-eq)	1.69	2.42
Total	(mg CO_2-eq)	185	295

Table 18: Energy Consumption for Different Modes of Transportation
(Per tonne-km Transported)

Truck	Tanker	Tank Car	Pipeline
kJ	kJ	kJ	kJ
1130	243	307	71.4

Table 19: Emissions Inventory for Transportation Scenarios 1, 3 & 4
(Per liter of FT Fuel Transported)

Transportation Mode		Truck	Tanker	Pipeline	Total
Southern Illinois to Chicago		DFO	RFO	Electricity	
Kilometers		97	0	322	419
CO_2	(mg)	7474	0	1321	8795
CH_4	(mg)	0.40	0	3.27	3.67
N_2O	(mg)	0.24	0	0.07	0.32
SOx	(mg)	36.7	0	12.86	49.6
NOx	(mg)	32.3	0	4.88	37.2
CO	(mg)	43.3	0	1.56	44.8
PM	(mg)	6.21	0	3.55	9.76
VOC	(mg)	0.28	0	0.04	0.32

Table 20: Emissions Inventory for Transportation Scenario 2
(Per liter of FT Fuel Transported)

Transportation Mode		Truck	Tanker	Pipeline	Total
Wyoming to Chicago		DFO	RFO	Electricity	
	Kilometers	97	0	1609	1706
CO_2	(mg)	7474	0	6605	14080
CH_4	(mg)	0.40	0	16.34	16.7
N_2O	(mg)	0.24	0	0.37	0.61
SOx	(mg)	36.7	0	64.3	101
NOx	(mg)	32.3	0	24.4	56.7
CO	(mg)	43.3	0	7.81	51.1
PM	(mg)	6.21	0	17.7	24.0
VOC	(mg)	0.28	0	0.18	0.46

Table 21: Emissions Inventory for Transportation Scenario 5
(Per liter of FT Transported)

Transportation Mode		Truck	Tanker	Pipeline	Total
Venezuela to Chicago		DFO	RFO	Electricity	
	Kilometers	97	3219	1931	5246
CO_2	(mg)	7474	57571	7926	72971
CH_4	(mg)	0.40	76.5	19.6	96.5
N_2O	(mg)	0.24	1.32	0.44	2.01
SOx	(mg)	36.7	723	77.2	836
NOx	(mg)	32.3	189	29.3	251
CO	(mg)	43.3	33.0	9.37	85.6
PM	(mg)	6.20	43.6	21.3	71.1
VOC	(mg)	0.28	28.5	0.21	30.0

Table 22: Emissions Inventory for Transportation Scenarios 6
(Per liter of FT Fuel Transported)

Transportation Mode		Truck	Tanker	Pipeline	Total
ANS to San Francisco		DFO	RFO	Electricity	
	Kilometers	97	6647	1287	8031
CO_2	(mg)	7474	118883	5284	131642
CH_4	(mg)	0.40	158	13.1	171
N_2O	(mg)	0.24	2.74	0.30	3.28
SOx	(mg)	36.7	1492	51.4	1580
NOx	(mg)	32.3	390	19.5	442
CO	(mg)	43.3	68.0	6.24	117
PM	(mg)	6.21	90.1	14.2	111
VOC	(mg)	0.28	58.8	0.14	29.2

Table 23: Greenhouse Gas Emissions from Transportation
(Per liter of FT Fuel Transported)

		Truck	Tanker	Pipeline	Total
Scenario 1, 3 & 4	(g CO_2-eq)	7.56	0	1.41	8.97
Scenario 2	(g CO_2-eq)	7.56	0	7.10	14.62
Scenario 5	(g CO_2-eq)	7.56	59.6	8.47	75.6
Scenario 6	(g CO_2-eq)	7.56	123.1	5.65	136.3

Table 24: Emissions Inventory for Ancillary Feedstocks

	Electricity	Diesel Truck	Heavy Equip.	Tanker	Fuel Gas	Butane	Methanol
	Delivered	Delivered & Consumed	Delivered & Consumed	Delivered & Consumed	Consumed	Delivered	Delivered
	(mg/MJ)	(mg/MJ)	(mg/MJ)	(g/MJ)	(g/MJ)	(mg/L)	(mg/L)
MJ/L		38.7	38.7	41.7			
CO_2	197500	76299	76299	82153	Calculated	162645	70269
CH_4	489	4.1	4.1	14.4	1.2	579	704
N_2O	11.7	2.5	1.9	1.9	1.9	5.3	10.0
SOx	1922	375	430	1031	0.0	50.9	642
NOx	500	330	888	775	60.3	937	1038
CO	56.9	442	383	287	14.6	218	238
VOC	503	88.3	64.8	144.1	2.6	1352	1415
PM	531	63.4	66.8	92.4	1.3	42.1	69.8

MJ = megajoule = 1e6 joules

Table 25: CO$_2$ Emissions from Combustion of Selected Fuels

FT Gasoline/Naphtha	Wt. % C	g CO$_2$/L
Design Option 1	85.63	2259
Design Option 2	85.05	2220
Design Option 3	78.73	2067
Design Option 4	85.63	2259
Design Option 5	86.81	2328
Design Option 6	85.95	2273
Design Options 7, 8	84.60	2129
FT Distillate		
Design Options 1, 2, 4-8	84.60	2381
Design Option 3	84.86	2366
	Wt. % C	**g CO$_2$/N liter**
Flared Associated Gas	61.96	2.09
	Wt. % C	**g CO$_2$-eq/N liter**
Vented Associated Gas	61.96	11.7

Table 26: Vehicle Fuel Economy-Technology Matrix
(Kilometers-per-liter)

Spark Ignition									
Conventional	4.3	6.4	8.5	10.6	12.8	14.9	17.0	19.1	21.3
Hybrid Electric	6.9	10.4	13.8	17.3	20.7	24.2	27.6	31.1	34.6
Direct Injection	5.4	8.1	10.8	13.4	16.2	18.8	21.5	24.2	26.9
Hybrid/Direct Inject	8.2	12.2	16.4	20.5	24.5	28.6	32.7	36.8	40.9
Compression Ignition									
Conventional	5.7	8.5	11.3	14.2	17.0	19.8	22.7	25.5	28.3
Advanced	6.5	9.8	13.0	16.3	19.6	22.8	26.1	29.3	32.6
Hybrid Electric	8.5	12.8	17.1	21.3	25.6	29.8	34.1	38.4	42.6
Advanced Hybrid	9.8	14.7	19.6	24.5	29.4	34.3	39.2	44.1	49.0

Table 27: Emissions Inventory for FT Fuels at Point of Sale
(Per liter of FT Fuel Supplied)

		Scenario 1	Scenario 2	Scenario 3	Scenario 4	Scenario 5	Scenario 6
CO_2	(g)	3395	3663	-1734	1119	1687	1746
CH_4	(g)	6.86	0.99	0.12	3.93	1.60	1.68
N_2O	(g)	0.02	0.02	0.17	0.02	0.02	0.03
SOx	(g)	1.54	2.27	0.05	0.06	0.84	1.58
NOx	(g)	0.66	0.88	4.72	3.08	2.84	3.03
CO	(g)	0.15	0.18	1.41	0.79	0.72	0.75
VOC	(g)	0.45	0.65	0.70	4.35	3.52	3.55
PM	(g)	0.39	0.62	0.08	0.02	0.08	0.12

Table 28: Full Life-Cycle GHG Emissions for FT Diesel
(g CO_2-eq/kilometer in SUV)

Scenario/ FT Feedstock Source	Extraction/ Production	Conversion/ Refining	Transport./ Distribution	End Use Combustion	Total Fuel Chain
1) IL #6 Coal	16	337	1	229	583
2) Wyoming Coal	4	364	1	229	598
3) Plantation Biomass*	-602	437	1	229	65
4) Pipeline Natural Gas	44	75	1	229	349
5) Venezuelan Assoc. Gas	32	132	7	229	399
6) ANS Associated Gas	32	132	13	229	405

Table 29: Full Life-Cycle GHG Emissions for Power Exported from FT Plants
(g CO_2-eq/MJ of Electric Power)

Scenario/ FT Plant Feedstock	All Upstream	Electricity Generation	Total Fuel Chain	Electric Efficiency
3) Plantation Biomass	-316	230	-86	60%
4) Pipeline Natural Gas	39	68	107	35%
6d) ANS Associated Gas	16	30	47	60%
U.S. Average All Plants	21	190	211	-
U.S. Average PC Plants	14	276	290	32%
NSPS PC Plant	13	255	268	35%
LEBS PC Plant	6	201	206	42%
Biomass Gasification Combine-Cycle	-237	247	11	37%

Table 30: Life-Cycle Sensitivity Analysis for FT Diesel
(g CO_2-eq/kilometer in SUV)

Scenario/ FT Feedstock Source	GHG Emissions Reduction		Total Fuel Chain	
			existing diesel engine	advanced diesel engine
1a) IL #6 coal - base case	-	-	583	507
1b) with seq. of process CO_2	279	48%	304	265
1c) with seq. of process & comb. CO_2	321	55%	263	229
1d) with co-prod. of fuels & power	189	32%	395	343
1e) with co-proc. of biomass	96	17%	487	424
1f) with coalbed CH_4 capture	14	2.3%	570	496
1g) with co-proc. of coalbed CH_4	145	25%	438	381
4a) Pipeline natural gas - base case	-	-	350	304
4b) with seq. of process CO_2	40	12%	309	268
4c) with seq. of process & comb. CO_2	75	22%	275	239
5a) Venezuelan assoc. gas - base case	-	-	400	347
5b) with flaring credit	359	90%	40	35
5c) with venting credit	2010	503%	-1611	-1401
6a) ANS associated gas _ base case	-	-	405	352
6b) with seq. of process CO_2	58	14%	347	301
6c) with seq. of process & comb. CO_2	131	32%	274	238
6d) with co-prod. of fuels & power	74	18%	332	288

Table 31: Full Life-Cycle GHG Emissions for Petroleum Diesel
(g CO_2-eq/kilometer in SUV)

Crude Oil Source	Extraction/ Production	Conversion/ Refining	Transport./ Distribution	End Use Combustion	Total Fuel Chain
Wyoming Sweet (40°API)	14	46	5	226	291
Canadian Light	19	50	7	228	304
Brent North Sea (38°)	14	50	5	228	298
Arab Light (38°)	22	50	16	228	316
Alaska North Slope (26°)	17	63	9	235	324
Alberta Syncrude (22°)	20	65	6	230	321
Venezuelan Heavy Oil (24°)	20	67	8	237	332
Venezuelan Syncrude (15°)	20	89	6	242	357